CASTEL DEL MONTE

Heinz Götze

CASTEL DEL MONTE

Geometric Marvel of the Middle Ages

Prestel

Munich · New York

For Wolfgang Götze
October 7, 1943–July 18, 1974

This is the expanded English edition of the third
German edition published in 1991.
Translated by Mary Schäfer

© 1998 Prestel-Verlag, Munich · New York

Front cover: Castel del Monte in the early morning sun;
view from the road leading from Andria
Back cover: Vector diagram of the basic geometric figure
of the ground plan of Castel del Monte
Frontispiece: The symmetrical octagonal inner court of Castel del Monte
viewed from the entrance level looking skyward

Prestel books are available worldwide.

Please contact your nearest bookseller or write to either of the following
addresses for information concerning your local distributor:
Prestel-Verlag, Mandlstrasse 26, D-80802 Munich, Germany
Tel.: +49-89-38 17 090; Fax: +49-89-38 17 09 35
e-mail: prestel@compuserve.com
and 16 West 22nd Street, New York, NY 10010, USA
Tel.: (212) 627-8199; Fax: (212) 627-9866

Die Deutsche Bibliothek – CIP-Einheitsaufnahme

Götze, Heinz: Castel del Monte: Geometric Marvel of the
Middle Ages/Heinz Götze. – Munich; New York: Prestel, 1998

Designed by Heinz Ross, Munich

Color separations by Karl Dörfel GmbH, Munich
Printed on acid-free paper "BVS Plus," 150 g/sm
Printed by Pera-Druck KG, Gräfelfing
Bound by R. Oldenbourg GmbH, Kirchheim

Printed in Germany

ISBN 3-7913-1930-2

Contents

Foreword to the English Edition	9
Foreword to the Third German Edition	12
Foreword to the First German Edition	15

I Introduction — 19

II Lines of Development of Hohenstaufen Architecture — 29
  1 Early History — 29
  2 Castel Maniace — 33
  3 Augusta — 43
  4 Castel Ursino — 44
  5 Lucera — 60
  6 The Gate to the Bridge of Capua — 67
  7 Caserta Vecchia — 72
  8 Prato — 74
  9 Termoli — 77
  10 Enna — 80
  11 Osterlant (by R. Spehr) — 83
  12 Castel del Monte — 89

III The Language of Forms in Hohenstaufen Architecture — 109

IV Castel del Monte: Design and Construction — 113
  1 The Octagon — 115
  2 The Eight-pointed Star — 129
  3 The Ground-plan Design: General Remarks — 140
  4 The Geometric Structure of the Plan — 147
  5 The Geometric Construction and Measurements — 158
  6 Spandrel Formation — 175

V The Geometric System and Its Realization — 183

VI Provenance of the Plan — 191

VII Interpretation — 206

Notes — 211
Selected Bibliography — 225
Sources of Illustrations — 232
Index of Names and Places — 235

Castel del Monte, Apulia: view from the east

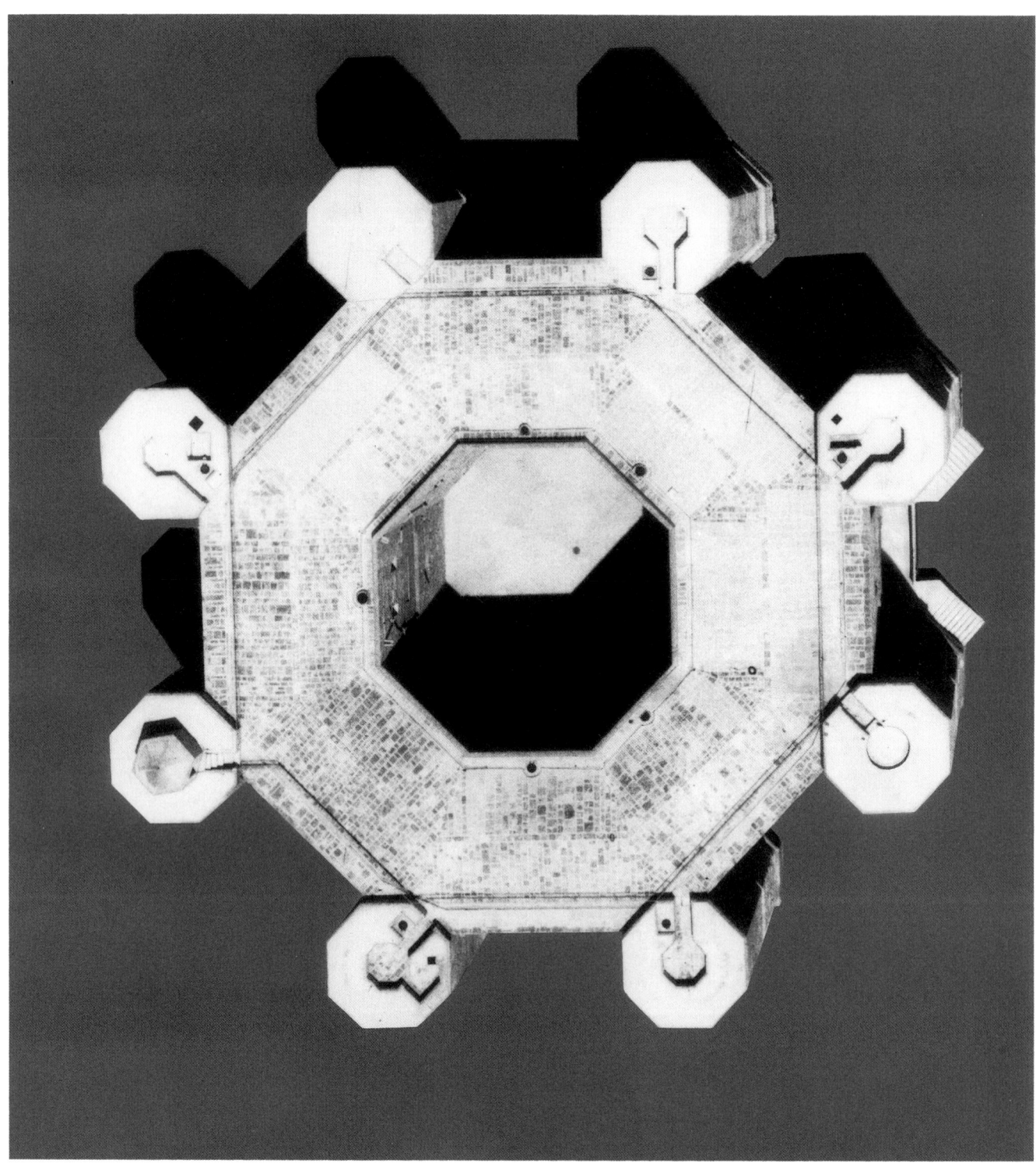

Aerial photograph of Castel del Monte. The geometric concept of the construction design becomes even more obvious here than from the normal view of the observer. The play of shadows accentuates the sharply contoured outlines of the similar, regular eight-sided figures, as do the exact right angles formed by the edges of the towers where they meet the main octagon.

Foreword
to the English Edition

I am deeply indebted to Prestel-Verlag for making possible an English edition of my book about Castel del Monte. Despite the favorable reception with which my book has met, some reactions indicate that the geometric prerequisites which are indispensible to understanding it are not always present. I am happy to have the opportunity to present herein further convincing evidence for the strictly geometric concept of the ground plan which also confirms its partly Islamic origin (see pp. 191 ff.). In particular, the outstanding publication of the Topkapi Scroll by Gülru Necipoğlu shows the connections that exist. The ornamental groups illustrated there offer interesting possibilities for comparison.

In the recent past I have taken pleasure in stimulating, enriching discussions with Yvonne Dold-Samplonius, Heidelberg; Heinz Halm, Tübingen; David Speiser, Basel; and Gunther Wolf, Heidelberg. As for the rest, what I wrote in the preceding forewords still holds.

Many thanks go to Prof. Marcel Berger of Paris, who dedicated his new edition of the classic work by D. Hilbert and S. Cohn-Vossen, *Anschauliche Geometrie*, to me and reproduced a computer simulation of Castel del Monte in the foreword.

I am especially obliged to Mr. Reinhard Spehr of Dresden, who let me have a description and pictures of Osterlant Castle near Oschatz (here, pp. 83–88). The castle was built in 1211/12 by the Margrave Dietrich von Meißen, a vassal of Frederick II.

Finally, I want to thank Mrs. Mary Schäfer for her outstanding translation, Mrs. Susan Siegle of Vienna for her knowledgeable recommendations, Mrs. Monika Passon for her technical assistance in diverse areas, and Ms. Christine Lackner for her excellent drawings. I am particularly grateful to the Messrs. Bernd Grossmann, New York, and Gerald Graham, London, for their willingness to do a critical reading of the English text.

Since 1989, when I secured the official permission of the *soprintendente*, architect Riccardo Mola, the working group of architectural historians under the direction of Prof. Dr. Wulf Schirmer, Karlsruhe, has carried out the surveying and recording of the monument Castel del Monte. It is hoped that — following some partial publications — a general presentation will soon appear within the framework of the Castel del Monte commission under the auspices of the Heidelberg Academy of Sciences, directed by Prof. Dr. Hans Elsässer. I am grateful to Prof. Dr. W. Schirmer and Dr. D. Sack for occasionally making information concerning selected measurement results available in advance.

Without making any claims to completeness, I would like to call attention here to some of the more recent literature:

Bauer, Friedrich L.: 'Sternpolygone und Hyperwürfel', in: *Miscellanea Mathematica*, pp. 7–43. Berlin, Heidelberg, New York 1991

Hell, G. and J. Otto: 'Photogrammetrische Arbeiten an Castel del Monte', in: *Allgemeine Vermessungsnachrichten* 4/92, pp. 169–174

Klein, U. and W. Zick: 'Castel del Monte — der geodätische Beitrag zur ersten präzisen Bauaufnahme', in: *Allgemeine Vermessungsnachrichten* 4/92, pp. 163–169

Koecher, Max: 'Castel del Monte und das Oktogon', in: *Miscellanea Mathematica*, pp. 221–223. Berlin, Heidelberg, New York 1991

Necipoğlu, Gülru: *The Topkapi Scroll — Geometry and Ornament in Islamic Architecture*. The Getty Center for the History of Art and the Humanities, Santa Monica 1995

Roshdi, Rashed: 'Fibonacci et les Mathématiques Arabes', in: *Micrologus II, Le scienze alla corte di Federico II*, pp. 145–160. Paris 1994

Sack, Dorothée and Wulf Schirmer: 'Castel del Monte', in: Koldewey Gesellschaft. Bericht über die 37. Tagung für Ausgrabungswissenschaft und Bauforschung, May 27–31, 1992 in Duderstadt, pp. 84–91

Sack, Dorothée: 'Castel del Monte e l'Oriente', in: *Federico II, immagine e potere*. Exhibition catalogue, Bari 1995, pp. 295–303

Schaller, Hans Martin: 'Die Kaiseridee Friedrichs II.', in: *Probleme um Friedrich II.*, pp. 109–134. Sigmaringen 1974

Schaller, Hans Martin: 'Die Staufer und Apulien', in: *Schriften zur staufischen Geschichte und Kunst*, vol. 13, pp. 125–142. Göppingen 1993

Schaller, Hans Martin: 'Die Frömmigkeit Friedrichs II.', in: *Schriften zur staufischen Geschichte und Kunst*, vol. 15, pp. 128–151. Göppingen 1996

Schaller, Hans Martin: 'Die Herrschaftszeichen Kaiser Friedrichs II.', in: *Schriften zur staufischen Geschichte und Kunst*, vol. 16, pp. 58–105. Göppingen 1997

Schirmer, Wulf with G. Hell, U. Hess, D. Sack and W. Zick: 'Castel del Monte. Neue Forschungen zur Architektur Friedrichs II.', in: *architectura* 1+2, 1994, pp. 185–224

Schirmer, Wulf: 'Castel del Monte', in: *Akademie Journal* 1/94, pp. 16–20

Schirmer, Wulf: 'Castel del Monte: Osservazioni sull'edificio', in: *Federico II, immagine e potere*. Exhibition catalogue, Bari 1995, pp. 285–294

Schirmer, Wulf and Dorothée Sack: 'Castel del Monte', in: *Kunst im Reich Kaiser Friedrichs II. von Hohenstaufen*, pp. 35–44. Munich, Berlin 1996

Stürner, Wolfgang: *Friedrich II., Teil I: Die Königsherrschaft in Sizilien und Deutschland 1194–1220*. Darmstadt 1992

Tits, Jacques: 'Symmetrie', in: *Miscellanea Mathematica*, pp. 293–304. Berlin, Heidelberg, New York 1991

Wolf, Gunther G.: 'Die Wiener Reichskrone', in: *Schriften des Kunsthistorischen Museums*, vol. 1. Vienna 1995

Wolf, Gunther G.: *Satura Mediaevalis, vol. III: Stauferzeit*. Heidelberg 1996

It should be noted that all Arabic place and proper names mentioned in the text are cited throughout in the customary scientific form and according to the recommendations for transcription from the German Oriental Society, even when one or the other term may be more familiar in English transcription due to its special popularity. A transcription table is therefore included on p. 210 as an aid to pronunciation.

Heidelberg, Spring 1998   HEINZ GÖTZE

Foreword
to the Third German Edition

Seven years have passed since the first edition of this book appeared. Among experts, interpretation of the historically unique monument Castel del Monte is still governed by the contradicting viewpoints concerning the origin of style and form in the architecture of Frederick II. One camp sees them as stemming from the spirit of central European Gothic, which evolved to classical form in the core areas of Burgundy and the Champagne. The doyen of research into the history of the art and architecture of Lower Italy — Émile Bertaux — championed this view. The other camp focuses on the unanswered questions raised by the relationship of the Hohenstaufen four-wing constructions to Syrian buildings.

It cannot be denied that the Arab rule over Sicily, which lasted for more than two centuries and had an influence that radiated into southern Italy, and the considerable Arab element that was consequently felt in the Norman world of the 'regnum', as well as Frederick II's ties with Arab culture and science, rooted in his Norman heritage, left their mark on the architecture of Lower Italy and Sicily. Wolfgang Krönig was the first to draw attention to the connection between Syrian buildings in the region of the 'limes arabicus' and the Hohenstaufen four-wing constructions.[1] Further discussion of this subject has become bogged down — for one thing, because a convincing line connecting Syria with Sicily and Lower Italy was missing, and for another, because further analysis of Frederick II's architecture embraced essentially the world of forms defined by the Cistercians, and the subject was thus considered almost exclusively from a central European standpoint. The same formal criteria were applied to Lower Italy and Sicily as were customary for central Europe, even though the strict geometric-mathematical forms and their systematics, perceptible everywhere, must have made the fundamental structural difference apparent. I have attempted to point out these mathematical-geometric qualities of Hohenstaufen architecture in southern Italy and Sicily and to tie them to the scientific history of that period. The great achievements of axiomatic Greek mathematics found their way via the kingdoms of the Diadochi into the Persian-Indian-Arab district. By way of the Islamic kingdoms, the stream of knowledge again reached southern Europe — Spain, France, and southern Italy. The introduction of Indo-Arabic numerals and the spread of the advanced knowledge of the great Arab mathematicians of the tenth and eleventh centuries in Europe by Leonardo Fibonacci (1170–1240 A.D.)[2] of Pisa characterizes this period.[3]

The field of geometry also contains the key to a broader understanding of the architecture of Castel del Monte in the sense of the ground-plan construction that I elucidated in the first edition; new sources confirm the correctness of this approach.

The absence of precise measurements and professional photographs of the building has up to now always been annoying.[4] Quite early on, therefore, I endeavored to remedy it. On May 22, 1989, the experienced and understanding *soprintendente* of Bari, architect Riccardo Mola, gave me written permission to have the monument surveyed and photographed. For this I am especially grateful to him. The Heidelberg Academy of Sciences, under its president Prof. Dr. med. Dr. med h.c. mult. Gotthard Schettler, kindly agreed to declare the task of surveying and professionally photographing an undertaking of the Academy. I am greatly obliged to the secretary of the Natural Sciences-Mathematics Section of the Academy and head of the Academy Commission, Prof. Dr. Hans Elsässer, for the realization of the Castel del Monte enterprise.

Most especially, I wish to thank the boards of trustees of the Robert Bosch Foundation and the Gerda Henkel Foundation, who, through generous donations, made it possible to carry out the survey, assisted by the architect R. Mola.

My active endeavor to incorporate in the work knowledge from neighboring disciplines has thankfully met with a large measure of understanding. Numerous friends and scholars have readily given their support. I feel exceptionally obliged to Mr. Heinz Halm in Tübingen, who helped me untiringly with information about the Arab-Islamic world, and to Mr. Marcel Erné of Hannover, who was my understanding advisor in the field of mathematics and contributed his own suggestions for solutions. My most sincere thanks are extended to both of them! In my studies of the geometric ornamentation of the Alhambra in Granada I received help and kind advice from Mr. José Maria Montesinos-Amilibia, professor at the Universidad Complutense in Madrid and member of the Spanish Academy of Sciences, and from Mr. Rafael Pérez-Gómez, professor of mathematics at the University of Granada. I am also grateful to the Messrs. J. van Ess and Alastair Northedge for valuable suggestions and pointers.

Heidelberg, Summer 1991   HEINZ GÖTZE

Foreword
to the First German Edition

Deep in the southern part of Italy, in the region of Apulia, stands one of the most beautiful fortresses of the Middle Ages, Castel del Monte, built by the Hohenstaufen emperor Frederick II. Various opinions exist concerning its significance, often of a speculative nature. The picture presented here is intended as an aid to arriving at a deeper understanding of the intellectual and architectural concept. In some respects, the singularity of the shape of Castel del Monte has confounded proven methods of research in art history. Other modes of access seemed necessary to do full justice to the ingenious concept. Clues to a new approach did exist: the 'crystalline' nature of the building with its symmetrical relationships has occasionally been pointed out but never systematically described and pursued. After all, since ancient times architects had been mathematicians! Presumably, even architecture itself — along with the art of surveying — was the mother of geometry. It is therefore necessary that in the interpretation and explanation of architectural masterpieces mathematics be considered more than it has been hitherto.

In the case of Castel del Monte, the regular octagon appears as the fundamental geometric design figure and in multiple symmetrical refractions it intensifies the overall aesthetic effect in the sense of Christian von Ehrenfels' 'Gestalthöhe' (height of form): of unity in diversity.

The book is also intended to draw attention to the plastic character of Hohenstaufen architecture in southern Italy, an architecture which exists in a three-dimensional mind that has repressed superficial ornamentation. This plastic sense is reminiscent of the classical Greek period and is visible in both architecture and sculpture. The pictures were chosen for the purpose of making this element of plasticity perceptible. The same goal is served by color close-ups, which attempt to catch the plastic-stereometric character of the buildings. The photographic sharpness of the individual shape was therefore subordinated to the optical effect as a whole. The choice of monuments illustrated was based primarily on architectural aspects.

The most recent summary (1974) of the literature which has appeared since 1904 was made by Wolfgang Krönig: *L'Art dans l'Italie méridionale*, Aggiornamento dell'Opera di Émile Bertaux sotto la Direzione di Adriano Prandi, vol. 5, Rome 1978, pp. 929–932. A critical characterization of the fundamental works is found in the noteworthy essay by Carl A. Willemsen, 'Die Bauten der Hohenstaufen in Süditalien. Neue Grabungs- und Forschungsergebnisse', in: *Arbeitsgemeinschaft für Forschung des Landes Nordrhein-Westfalen, Geisteswissenschaften*, no. 149, 25th issue, Cologne/Opladen 1968, p. 9 ff.

I am grateful to numerous friends and scholars for bibliographical hints and a good deal of advice. I would like to name in particular Friedrich Wilhelm Deichmann, Rome, Hans Junecke, Berlin, and Max Koecher(†), Münster/Westfalen, in addition to Albrecht Dihle, Wilhelm Doerr, Yvonne Dold, Erwin Walter Palm(†), Peter Anselm Riedl, Dietrich Seckel, all of Heidelberg, Hans Martin Schaller, Munich, and Carl A. Willemsen(†), Bonn. I am especially grateful to Mr. Heinz Sarkowski of Heidelberg for his critical reading of the proofs.

Moreover, I thank the branches of the German Archaeological Institute in Istanbul (Wolfgang Müller-Wiener), Madrid (H. Schubart), and Rome (Theodor Kraus), the *soprintendenza* in Bari, in Palermo, and in Syracuse, and finally the Photography Department of the Bibliotheca Hertziana in Rome, Ms. Eva Stahn.

Through the generosity of the secretary of the Heidelberg Academy of Sciences, Philosophical-Historical Section, Hans-Joachim Zimmermann, I had the privilege of presenting there my studies on the ground-plan design of Castel del Monte.

Finally, my thanks go to Prestel-Verlag for providing the excellent makeup of this book.

Heidelberg, Whitsuntide 1984   HEINZ GÖTZE

General map of the Kingdom of the Two Sicilies in the mid-thirteenth century. Places mentioned in the text are underlined.

1 Portrait bust, taken to be a likeness of Emperor Frederick II. Museo Civico, Barletta, Apulia.

I

Introduction

Grandly did the Orient make its way
across the Mediterranean;
Only one who loves and knows Hafez
understands whereof Calderón sang.

Johann Wolfgang von Goethe,
Westöstlicher Diwan, Buch der Sprüche[5]

The character of Frederick II, variously portrayed from a standpoint of favor or contempt, has found renewed interest in recent years — not only in Germany, but also particularly in France, where research into the history and art of southern Italy during the Middle Ages commenced even earlier.

A study of the architecture of the Hohenstaufen period is inconceivable without an idea of the emperor as a person. It is not easy, however, to obtain a reliable picture of this truly brilliant ruler, who practically challenged chroniclers to depict him with a bias to one side or the other.

It is difficult to determine how Frederick II appeared outwardly. The 'Augustan' coins (Fig. 2) present an idealized portrait after the likeness of Augustus Caesar. Salimbene, a Franciscan Minorite who saw the emperor in person in 1238, described his appearance as follows: "Pulcher homo et bene formatus, sed medie stature fuit."[6]

In the Vatican Library there is a recently discovered drawing (Fig. 3) of the statue of the seated emperor that stood at the gate to the bridge of Capua (p. 69, Fig. 96), the head of which is said to have been thrown into the Volturno River by French revolutionary troops in 1799. With the help of this drawing, P.C. Claussen has attempted to establish that a terracotta bust by Tommaso Solari (d. 1779) in the Campano Museum is a copy of the missing plaster of paris cast, or even of the original.[7] The head, however, is not very expressive (see Fig. 4).

Two large sculptures are regarded as contemporary portraits of the emperor, but they cannot be assigned with certainty to that period. The most impressive is the remains of a bust discovered in 1953 by Adriano Prandi in the storerooms of the museum at Barletta; the head is shown in Fig. 1. At the base of the bust is a partially de-

2 Augustan coin depicting Frederick II of Hohenstaufen. Stadtarchiv Göppingen.

19

stroyed inscription that experts date much later than the Hohenstaufen period. Numerous factors indicate that the bust was the upper part of a larger than life-sized statue of a horseman with a height of approximately 3.20 m (the Bamberg horseman is 2.40 m!). This origin would explain the uneven finish of the base, which, along with the lowest part of the bust, shows fewer signs of corrosion and damage than the remainder of the bust and head, suggesting that the lower portions were reworked at a later date. The elongated neck indicates that the sculpture was designed to be seen from below and that it may well have been an equestrian statue. Careful stylistic analyses point to a date in the mid-thirteenth century. The bust would thus be a likeness of the emperor in the final years of his life. The appeal of the portrait lies in the contrast between the noble, majestic posture and the naturalistic depiction

3 Drawing of the seated figure of Frederick II from the bridge gate in Capua (cf. the draped torso, Fig. 96). The drawing stems from the estate of Séroux d'Agincourt. Rome, Vatican Library.

4 Bust of Frederick II, cast by Tommaso Solari. Museo Campano, Capua, Campania.

6 (p. 21, bottom)
Picture of Henry VI from the Manesse manuscript (Heidelberg, University Library, Cod. Pal. Germ 848, fol. 9 r.). The Hohenstaufen claim to the Sicilian kingdom of the Normans was realized with the marriage of Emperor Henry VI (1165–1197) to Constance, heiress to the Norman southern kingdom, in 1186.

5 Palermo, Sicily: Cathedral, sarcophagus of Emperor Frederick II.

of an ageing man with deeply furrowed crow's-feet, tear sacs that stand out in relief, and a certain flabbiness of form that can be seen in the eyelids, cheeks, and wrinkled forehead. The head is unfortunately disfigured as a result of weathering and deliberate destruction. Also badly damaged are the leaves of the wreath, which, in connection with the monumental figure, suggest that the subject was an imperial personage.

Even today — and particularly within the boundaries of the old kingdom of the Two Sicilies — the memory of Frederick II is amazingly alive. At the emperor's marble sarcophagus, which stands next to those of his first wife Constance, his father Henry VI, his grandfather Roger II, and the latter's daughter Constance in the cathedral of Palermo (Fig. 5), one almost always sees fresh flowers — more than 700 years after his death! And in Iesi, the town where he was born, the story of his birth is still known to everyone. The ancestral castle of the Hohenstaufen line lies in Germany. The likeness of Henry VI, son of Barbarossa and father of Frederick II, appears at the beginning of the Codex Manesse (Fig. 6), which originated between 1300 and 1340 and is now in the library of the University of Heidelberg.

Frederick was born on December 26, 1194, in the small village of Iesi, 25 km southwest of Ancona, as the son of Henry VI and his wife Constance, heiress to the Norman southern kingdom of the Hauteville dynasty. He was given the names Frederick and Roger, in memory of his grandfathers Frederick Barbarossa and Roger II. He lost his mother on November 26, 1198, and several months later, on May 17 (Whitsun), he was crowned king of Sicily. Frederick grew up in Sicily's capital city as an orphan under the guardianship of Pope Innocent III. There he was forced at an early age to prove himself and to develop his abilities in dealing with people. With its exciting multicultural splendor, the city of Palermo, chosen by the Arabs as their capital and magnificently endowed by the Normans, shaped the mind of the open, enthusiastic young boy who was receptive to everything of beauty. He was fascinated by the manifold impressions that presented themselves to him, and especially by the lifestyle and culture of the Arabs, which suited his intellectual curiosity. He was able to acquire the most comprehensive education of his time, because a spirit of openness to all

Introduction 21

points of view and the tolerance that developed therefrom had reigned at the court of his maternal ancestors.

Palermo was then the largest city in Europe; it maintained trade and travel relations with the entire Mediterranean world, and it is easy to imagine its variegated population. Founded by the Phoenicians in the seventh century B.C., it was never colonized by the Greeks. As the main base of the Carthaginians in Sicily, it was conquered by the Romans in 254 B.C., falling later to the Ostrogoths, who lost the city in turn to Belisarius in 535 A.D. Until its conquest by the Saracens in 830/31, Palermo remained Byzantine. The geographer Ibn Ḥauqal (tenth century) reported that there were hundreds of mosques in Palermo, more than he had seen anywhere else, with the exception of Córdoba.[8] After 1072, the Normans adopted the orderly administrative system of the Arabs, many of whom remained in the city. One of the young Frederick's tutors is said to have been the cadi, or judge, of the Muslim community in Palermo.[9] The emperor's command of Arabic and his knowledge of Arabic culture speak for the truth of this tale. It should be kept in mind that the Arabic-Norman Capella Palatina (1131/41) and San Giovanni degli Eremiti (1132; Fig. 7) were built only about 50 years prior to the emperor's birth, the 'Martorana' (1143; Fig. 11) and San Cataldo (1161; Figs. 8–10) just a few years later. It is quite possible that the muezzin's chanting was still to be heard in Frederick's youth. These structures presuppose the work of Arab-Saracen artists during the Norman period. We may also assume with certainty that purely Arabic

7 Palermo, Sicily:
San Giovanni degli Eremiti (1132).

8 Palermo, Sicily:
Church of San Cataldo (1161).

9 Palermo, Sicily:
Inside view of San Cataldo (1161), looking upward into two of the three cupolas.

10 Palermo, Sicily: Detail of the window in the apse of San Cataldo (1161). Arab grating pattern with the eight-pointed star motif.

edifices and fortifications existed during Frederick's early years.[10]

At the Norman court of William II, builder of the Cathedral of Monreale (Fig. 12), Ptolemy's *Almagest* was translated from Arabic into Latin by Byzantine scholars, at the same time as his writings on astronomy were being translated in Toledo. Also a part of this intellectual world were the works of Hippocrates, which may have paved the way for the advancement of the medical school of Salerno, founded by the Normans in the eleventh century.[11] Frederick's fondness for the flourishing Islamic culture of his time was rooted in this Palermo environment. It remained undiminished throughout his life. When his sarcophagus was opened in Palermo he was found to be not only dressed in his robes of state, but also wrapped in Arabic garments with Kufic characters.[12] During those early years in Palermo, however, Frederick also directly experienced intrigue, deception, and disloyalty. This strengthened the boy's sense of justice as the foundation for all personal and public life. At the same time, it may also have been the root of the distrust, of a certain cynicism, even, that often manifested itself in Frederick's life.

Introduction 23

In December 1208 he took over the business of governing himself, and in 1212 he set out to cross the Alps, heading north with a small company of knights; in the same year, in Nuremberg, he was elected to be the German king, and in 1215 he received the imperial crown in Aachen. His succession to the heritage of Charlemagne and of his grandfather Frederick Barbarossa made an indelible impression and remained of enduring importance for him from that time onward. He understood the power transferred — the empire bestowed by God — as an obligation to enforce a principle of order that would guarantee to the people under his rule personal safety under the protection of the imperial power, a secure peace, i.e., the 'Pax Augusta', and the right to an existence worthy of a human being. The concept of the 'Pax Augusta', of freedom in justice, ran through the emperor's life like a political leitmotif.

What was extraordinary about Frederick? What made him in his own time the 'wonder of the world' (*stupor mundi*), the 'hammer of the earth'? Where blood was concerned, he embodied not only the heritage of his Hohenstaufen ancestors, who were noted for their strong will and shrewd diplomacy, but also the bold determination to conquer and the political realism of the Norman Hautevilles, among whom were such figures as Robert Guiscard and the brilliant statesman Roger I. The combination of vitality with the faculty of critical judgement, of intuitive comprehension and resourcefulness, imparted to him splendor and superiority. Even his enemies were occasionally unable to suppress words of great admiration.

His outstanding achievement as a statesman was to complete the codification of the law in the constitutions of Melfi in August 1231, in which he was personally involved; it was drafted in the face of fierce hostility on the part of Pope Gregory IX, anticipating the latter's new codification of canon law that appeared in 1234.[13] The 'Constitutiones Melfitanae' represented the first comprehensive constitutional order in Europe, governed by a systematic basic concept expressed in a principle coined by the emperor: "Imperator est pater et filius iustitiae."[14] He, the emperor, establishes the law, to which at the same time he is subject like any other citizen of his kingdom. On this legal basis it was possible to set up a tightly

11 Palermo, Sicily: Campanile of the church of La Martorana (1143).

24   Introduction

12 Monreale near Palermo, Sicily: Exterior view of the apses of the cathedral, built by the Norman King William II beginning in 1174 in late Norman style. It is famous for its Romanesque bronze doors by Barisanus of Trani and Bonanus of Pisa.

organized bureaucracy against the background of the Arab-Norman administrative tradition. The equality of status before the law granted to everyone, be they Christian, Arab, or Jewish, and the religious freedom that went along with it are surprisingly modern and tolerant features of this legal establishment that corresponded to the Norman tradition of law and administration.

In the same spirit, Frederick had founded the University of Naples in July 1224, Europe's first state university, which retained its original legal form until 1812.[15] It was exclusively for the study of the sciences, whose patron Frederick considered himself — "vir inquisitor et sapientiae amator." Every kind of clerical influence, which predominated at the other large universities of that period, was precluded. The conferment of academic degrees and the prerequisite examinations were incumbent upon the grand chancellor of the empire himself or his representatives. Resolutely pursuing the concept of the regulatory power of the emperor in the sense of the Augustan em-

Introduction   25

pire, he lent weight to the doctrine of Roman law here in Naples. From the beginning, the university consisted of four faculties and was intended above all to train highly qualified, scientifically educated civil servants for administrative functions. From Roffredo de Benevento, one of the first important teachers of law, Andrea d'Isernia, and Saint Thomas Aquinas to Francesco de Sanctis, the University of Naples has boasted brilliant names among its professors.

In Salerno Frederick encouraged the highly reputed medical school to take charge of matters concerning public health and their organization in the kingdom of Sicily. The emperor was very concerned about establishing healthy, hygienic conditions. The provision that seriously ill persons without means were treated at no cost is evidence of social welfare, as is the fact that, under the 'Constitutiones Melfitanae', widows and orphans were able to seek counsel from the court without charge.

By a stroke of luck, a remarkable manifestation of Frederick's own scientific achievement has survived in full: the book of falcons, *De arte venandi cum avibus*,[16] still quoted today by ornithologists. By no means does it deal only with training falcons for the hunt; it also provides a systematic taxonomic survey of the various types of falcons and their way of life (Fig. 13).

The emperor is said to have had a good command of seven languages. He had a particular liking for mathematics, and he followed its development with great attention. He maintained close contact with the most eminent European mathematician of his time, Leonardo Fibonacci of Pisa (ca. 1170–ca. 1240),[17] the son of a secretary of the republic of Pisa who was entrusted with the management of a Pisan trading post in Bejaïa, Algeria, in approximately 1192. Pisa was then the most important trading city in the western Mediterranean, comparable only to Venice on the Adriatic. Fibonacci traveled on his father's behalf to Egypt, Syria, and Byzantium. Around the turn of the century, he went back to Pisa in order to devote himself exclusively to the study of mathematics. His greatest service was to impart the achievements of Arabic mathematics to the Western world. He introduced Indo-Arabic numerals and calculation with the symbols 0 (zero) and X to Europe. With the 'Fibonacci sequence' he created a mathematical expression for the harmonic ratio of the

13 Emperor Frederick II with a falcon. Miniature from the Book of Falcons by Frederick II. Bibliotheca Palatina in the Vatican Library.

golden section, already known to the Greeks. Leonardo was probably introduced to Frederick II in approximately 1225. The emperor prompted the scholar to revise his *Liber Abaci*, which had appeared in 1202; the revision was completed in 1228. In turn, Leonardo Fibonacci dedicated his work *Liber Quadratorum* to the emperor.

Finally, Frederick II founded the Sicilian school of poets. This was inspired by the minnesong of the Provence, which had probably found its way to Sicily with the emperor's first wife, Constance of Aragon (1209).[18]

What a wealth of intellect! And this in the face of endless political and military battles with the papal court and its allies. The Curia claimed primacy not only in spiritual but also in secular matters, and thus had already become a fundamental and irreconcilable opponent of the imperial concept regarding the regulatory power of the worldly empire under Frederick Barbarossa.

The word 'Antichrist' was a battle cry in this mounting dispute. Frederick lived entirely according to Christian tradition and was convinced that imperial power was a mandate from God. In a letter composed in magnificent style, following his coronation in Aachen, Frederick wrote to the general chapter of the Cistercian abbots:

"As We, though We are sinners, have been entrusted through God's unspeakable mercy with the helm of the Roman Empire, may He Himself, through your pious intercession, bestow upon Us the spirit of justice and truth, so that under Us the empire be governed and regulated such that, to the praise and glory of His name, His holy church may delight in the welcome repose of peace and We, when this temporal kingdom has run its course, may attain with you the kingdom that will have no end."[19]

At the same time, Frederick remained a continually questioning and doubting mind who did not accept the Church's dictates about what to believe and how to think. An unquenchable thirst for knowledge drove him and inspired the most astonishing experiments, which reflect none of the prejudices of medieval thought. A long letter to Michael Scotus has become famous, in which he posed a wealth of questions related to the natural sciences.[20]

His political behavior was unswervingly determined by his widespread goals, and he was inventive in choosing the means to attain them. While he could be very enigmatic in small matters, he was endowed with great diplomatic skill and statesmanlike wisdom, but his effusive temperament frequently got in the way of success.

Frederick's connections with the Cistercian order were lifelong, and it was his wish to be buried in the cathedral of Palermo wearing Cistercian robes. This religious order, which had particular ties to the traditions of medieval architecture, played a significant role in the design of the structures built by Frederick II.

The architecture, to which we will devote ourselves below, will show the emperor's creative genius in a dazzling light. This architecture proceeded from the soil of his German dominions and the French Gothic, but above all from that of his homeland in southern Italy, with its varied roots in the Greek, the Arab, and the Norman worlds. His genius pointed it, similar to the sciences and the art of governing, in new directions and allowed it to find new forms. The mathematical heritage of the ancients and of Islam, the plastic sense of the Greeks, and the early Gothic structures of central Europe all influenced it.

If, in spite of the dramatic termination of Frederick II's reign and the end of the Hohenstaufen dynasty that rapidly followed, his personality can be felt so strongly even today, how powerful an effect must this emperor have had on his contemporaries, who described him as "…the greatest among the princes of the world, Frederick, also the marvel of the world and its wondrous transformer!"[21] With his death in 1250, his attempt to restore the Augustan empire, undertaken with such energetic strength and persistence, had failed, and the idea of Western unity lost its fascination.

"His bitter struggle against the Church's claim to secular power nevertheless contributed considerably to the dissolution of the medieval world and thus to its emergence into modern times … The sense of reality became keener, and the beginnings of a new art and a new perception of nature became visible: beginnings of the world in which we live today."[22]

II

Lines of Development of Hohenstaufen Architecture

1 Early History

Frederick II developed into a tireless, productive builder in the kingdom of Sicily; his fortresses, citadels, and palaces number more than 200, but there are only a few churches, such as the cathedral of Altamura.[23]

Because of the large state expenditures it entailed, the incessant building worried even his most loyal followers. If we visualize the position in which Frederick II found himself at the beginning of his rule, it becomes quite clear that his construction zeal cannot be explained purely by the pleasure of building. He had to reckon with opponents who threatened from the sea as well as on land. He conducted a war on several fronts, the one at home being against the power and influence of the feudal aristocracy and for the establishment and maintenance of constitutional order.

In planning the chain of citadels — on the Adriatic and the eastern coast of Sicily, for example, but also in the interior — he first had to enlarge existing castles; depending on the circumstances, they were reconstructed or added to in accordance with the dictates of contemporary fortress achitecture, as for instance the ancient Messapian fortress of Oria in southern Apulia (Fig. 14). At the same time, new fortresses arose. It is often difficult to date them precisely. Many go back to Byzantine, Norman, and presumably also Islamic layouts. Even the early ones were clearly planned constructions with a tendency to plastic-stereometric form — for example, in Melfi (Fig. 15), or in the magnificent harbor fortress of Trani[24] (Figs. 16 and 19). The plans remind one strikingly of buildings from the Persian-Achaemenian period in Pasargadae and Persepolis

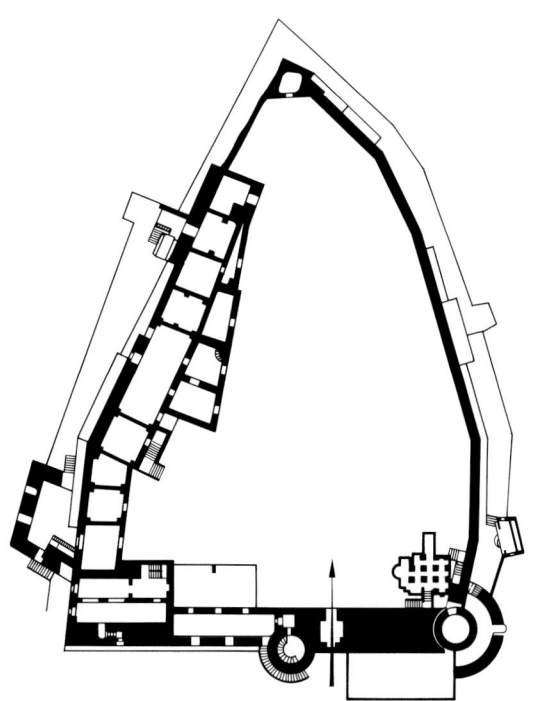

14 Oria, Apulia: Ground plan of the citadel (after Ceschi, with additions).

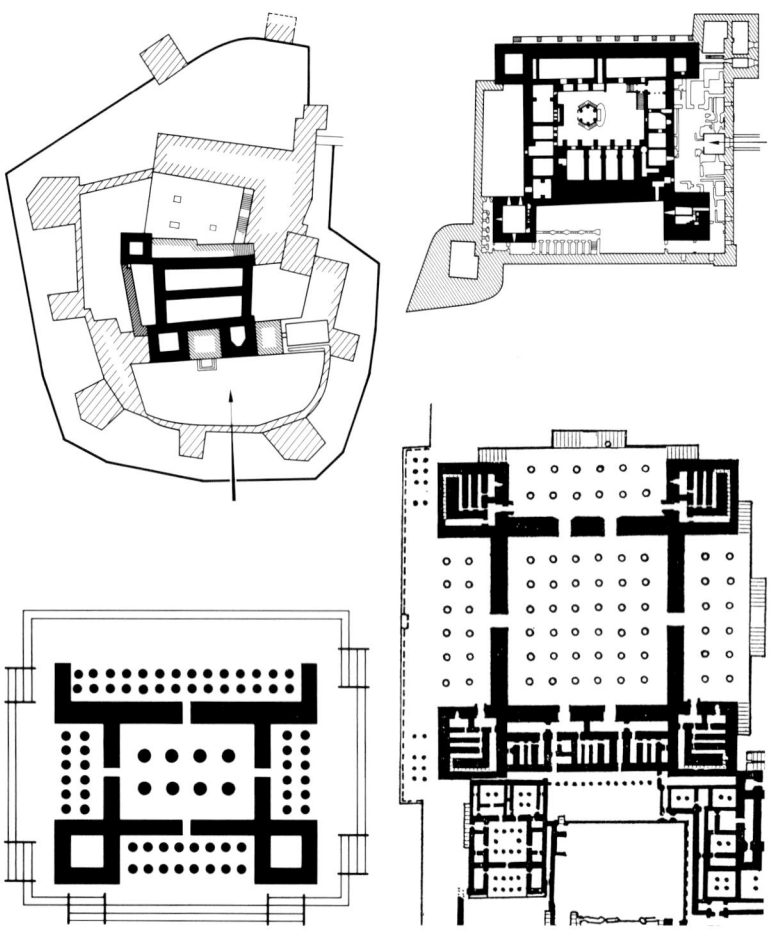

15 Melfi, Basilicata:
Ground plan of the citadel (from Lenz). Black outlines indicate the Hohenstaufen core of the building.

16 Trani, Apulia:
Ground plan of the port citadel (from C. A. Willemsen); post-Hohenstaufen structures are hatched.

17 Pasargadae, Iran:
Achaemenian structure from the middle of the sixth century B.C. (after a drawing by H. Frankfort).

18 Persepolis, Iran:
Plan of an Achaemenian building dating from the turn of the sixth to the fifth century B.C. (after a drawing by E. F. Schmidt).

(Figs. 17 and 18). In connection with Hohenstaufen halls — in particular Castel Maniace in Syracuse — R. Wagner-Rieger refers to Persian royal halls of the Achaemenids in Persepolis and, as intermediate stages, to Byzantine cistern constructions. He regards the colonnades and halls of Cistercian cloisters as the direct prototypes, however.[25]

There are precursors to the clearly recognizable plastic-stereometric design of Hohenstaufen architecture — for instance, in Norman architecture, with its enormous donjons at Adranò and Paternò (eleventh and twelfth centuries; see p. 62, Figs. 81 and 82) built on the remains of Saracen, perhaps even Greek structures. Consider also such impressive monuments in the region of Apulia as the now abandoned church of Ognissanti (Fig. 20) in Valenzano near Bari, with its severe stereometric shapes and square pyramids over the cupolas. The emperor's ever-open mind surely received stimulating and enriching impressions in discussions with architects and during travels and military campaigns and shaped them into new ideas. In Sicily,

30 *Lines of Development of Hohenstaufen Architecture*

19 Trani, Apulia:
View of the citadel from the east.

as in Lower Italy, however, he encountered important Norman and Roman architectural monuments, such as the splendid cathedral of Troia, decorated with orientalizing elements (p. 91, Fig. 130) or Santa Maria Maggiore di Siponto (p. 65, Fig. 88). There is much evidence for Frederick's interest in architecture. It was not exhausted in the building of strongholds; rather, his especially pronounced appreciation for aesthetics, for mathematics, and for classical art was no doubt combined with the awareness of the great political-symbolic significance of architectonic

20 Valenzano near Bari, Apulia:
The church Ognissanti di Cuti in an abandoned Benedictine abbey from the third quarter of the eleventh century.

Early History 31

21 Foggia, Apulia: Preserved portal arch of the otherwise totally destroyed palace of Emperor Frederick II.

concepts. It is soon apparent that the preconceived architectonic idea became the dominating element of his architecture. This becomes evident if we compare, for example, the plans of central European fortresses, which generally fit into the terrain, with Frederick's citadels, whose layouts are forced upon the landscape (cf. p. 34, Fig. 23).

The monumental record shows that Frederick's building activity served first of all external and internal security; alongside this there will have been a need for suitable residences and places to stay for hunting and recreation. We know how much he loved Foggia in the Capitanata. Unfortunately, all that remains of this originally particularly richly appointed palace are an inscription and an arch; the latter, with its large scale and the attention given to its workmanship, gives an idea of how magnificent it once must have been (Fig. 21).

It is amazing to see how, following the emperor's return from the crusade (1228/29), a group of important structures arose that, despite their varying design, are indicative of *one* great concept to which the emperor himself no doubt contributed. These structures that were built between 1230 and 1250 are directional for the systematic consideration and analysis of the underlying construction ideas and forms of expression. Among them, first of all, are three citadels on the eastern coast of Sicily that are related with respect to their structural plans, and that were started during the ten-year period beginning in

32   *Lines of Development of Hohenstaufen Architecture*

approximately 1235. All three are still standing today, albeit in varying degrees of preservation. The emperor had charged his 'praepositus aedificiorum', Ricardo da Lentini, with completing the three buildings, as indicated in a letter sent from Lodi, dated November 17, 1239.[26]

2 Castel Maniace

The oldest and probably most lovely of the three citadels is in Syracuse — Castel Maniace, begun prior to 1239.[27] It lies on the point of the Ortygia peninsula and is difficult to approach, which may have contributed to the fact that it has not yet received the appreciation it deserves as a true masterpiece of Hohenstaufen architecture (Figs. 22 and 24). The plan (Fig. 25) is strictly square, measuring 50×50 m, with cylindrical corner towers in which are mounted cylindrical spiral staircases encompassing *sixteen* steps.[28]

22 Syracuse, Sicily: Castel Maniace from the west.

In contrast to the other two citadels, the four wings are not broken up by side towers — rather, the walls extend from one corner tower to the next in squared-stone masonry, right down to the foundations. The bases of the four cylindrical corner towers change into conical frustums (Figs. 26 and 27). The absence of towers on the sides allowed the possibility of positioning the magnificent, splendidly appointed portal in the median axis of the front (Figs. 28–30).[29] This emphasizes the rigorously symmetrical form and the material solidity of the structure. On each side of the deeply graduated portal, at the height of the arches, once lay two Hellenistic bronze rams on consoles (Fig. 31). One survives and is now kept in the museum of Palermo. Frederick's great liking for classical sculpture is here as evident as is his taste for grandiose architecture.

Castel Maniace has four symmetrical axes, and the interior is divided into $5 \times 5 = 25$ squares covered by ribbed vaults. The network of vaults symmetrically spans the interior area. Eight inner vaulted squares join the corresponding outer squares on each side (16 altogether), resulting in 24 squares with pointed arches and solid buttresses, resting on 16 columns and 20 pilasters. The 25th square in the middle is left open and forms an inner courtyard

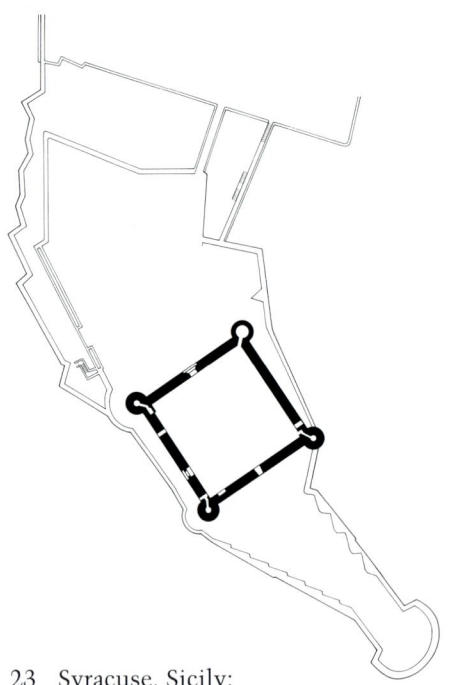

23 Syracuse, Sicily:
Layout of Castel Maniace on the outermost tip of the Ortygia peninsula.

24 Syracuse, Sicily:
Castel Maniace from the east.

34   *Lines of Development of Hohenstaufen Architecture*

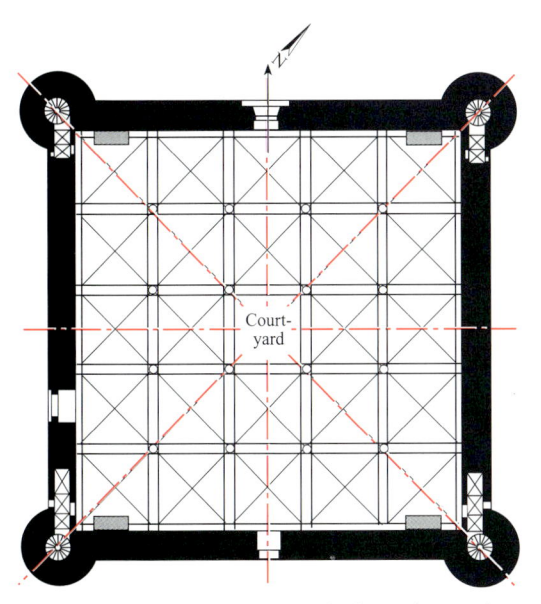

25 Syracuse, Sicily: Ground plan of Castel Maniace, with lines of symmetry in red.

26 Syracuse, Sicily: Southwest corner tower of Castel Maniace with conical base showing truncated surfaces.

27 Syracuse, Sicily: South corner tower of Castel Maniace. The tower contour is visible down to the base as a result of storm damage to the surrounding wall.

through both stories of the original building. The supports for this *impluvium*-like open center square were fashioned, with particular opulence, of three bundled half columns each.[30] They are strikingly related to the corresponding half columns in the upper story of Castel del Monte (pp. 110/111, Figs. 156 and 157). Here, as there, light marble with dark veining and uniformly dark marble was used, as careful restoration shows (Figs. 43 and 44). The bases of these column bundles are round (Fig. 45). The splendidly carved capitals are also related to those of Castel del Monte. They reveal the hand of great master craftsmen and are entirely on a par with the capitals and statuary of Castel del Monte (Figs. 38–41).

This courtyard, which let light into the rooms, is related in its motif to the inner court of the fortified castle of Lucera (p. 63, Fig. 85) but, in the last analysis, also to the *octagonal* court of Castel del Monte (p. 89, Fig. 125), in the middle of which there may have been a cooling fountain.[31] The columns and half columns stand on octagonal bases.[32] Only the slender bundled columns of the middle court rest on round bases.

The structure as a whole is impressive because of its stereometrically formed blocks of masonry, whose cut-

Castel Maniace   35

ting planes of completely unembellished light marble present their full solidity to the open space. Even the windows — as well as the portal — are not just breaks in the wall space but rather 'plastic instrusions'. The window on the southwest side is recessed into the wall in deeply inset steps (Fig. 35). Through the free-standing columns and the use of varicolored stones in the window jambs, the window opening becomes a three-dimensional ornament.[33]

All of this repeats itself in even grander form in the splendid main portal, already described (Figs. 28 and 30). Here, in the middle of the capital of 'Magna Graecia', the Greek appreciation for the descriptive power of a building was reborn — if indeed it had ever been lost. It is expressed in the use of squared stones in the construction from top to bottom (Fig. 37). The smoothly chiseled limestone blocks are carefully laid and present a picture that strikes one as classical — the spirit of Greek genius. The place certainly influenced the choice of material and technique: Since antiquity, Syracuse had obtained its fine limestone from its own well-known quarries, the *latomie*. Many of the Greek structures built with this stone were probably

28 Syracuse, Sicily: Detail of the portal of Castel Maniace.

29 Syracuse, Sicily: Portal side (northwest side) of Castel Maniace.

36   *Lines of Development of Hohenstaufen Architecture*

30   Syracuse, Sicily: Main portal on the northwest side of Castel Maniace.

31   One of the two early Hellenistic bronze rams (dimensions: 1.31 × 0.82 m) that stood on consoles to the right and left, over the portal of Castel Maniace in Syracuse. Museo Nazionale, Palermo, Sicily.

32   Syracuse, Sicily: View of the portal wall of Castel Maniace from the interior.

in far better condition during the time of Frederick II than they are today. It is even possible that the traditional skill of stone masonry had survived from ancient times. A classical understanding of construction and material is visible here that can also be felt in other Hohenstaufen structures, for instance, at the gate to the bridge of Capua,[34] in Enna, and finally in Castel del Monte.

Stepping through the portal, we are surrounded by an impression of the mighty ribbed groined vaulting (Fig. 42) that is difficult to put into words. Two of the five rows of vaults have been restored. One is able to imagine a truly royal hall. The plastically arranged capitals and imposts, the strongly accentuated ribs, the feeling of tremendous weight conveyed by the powerful arches lead less to a spatial than to a directly physical experience of these vaults.

Castel Maniace   37

33   Syracuse, Sicily:
Window lintel with relieving arch on the southeast side of Castel Maniace.

In its entirety it breathes a grandeur that avails itself of basic Gothic forms but radiates classical solidity. The question of whether the columned hall displays an Islamic influence or whether a Gothic sense of space predominates is answered by the realization that a classical appearance is evoked here with Gothic morphological elements. There is no connection with the columned halls of the Omayyad mosque in Córdoba. They do not convey the physically perceptible, overwhelming sense of space

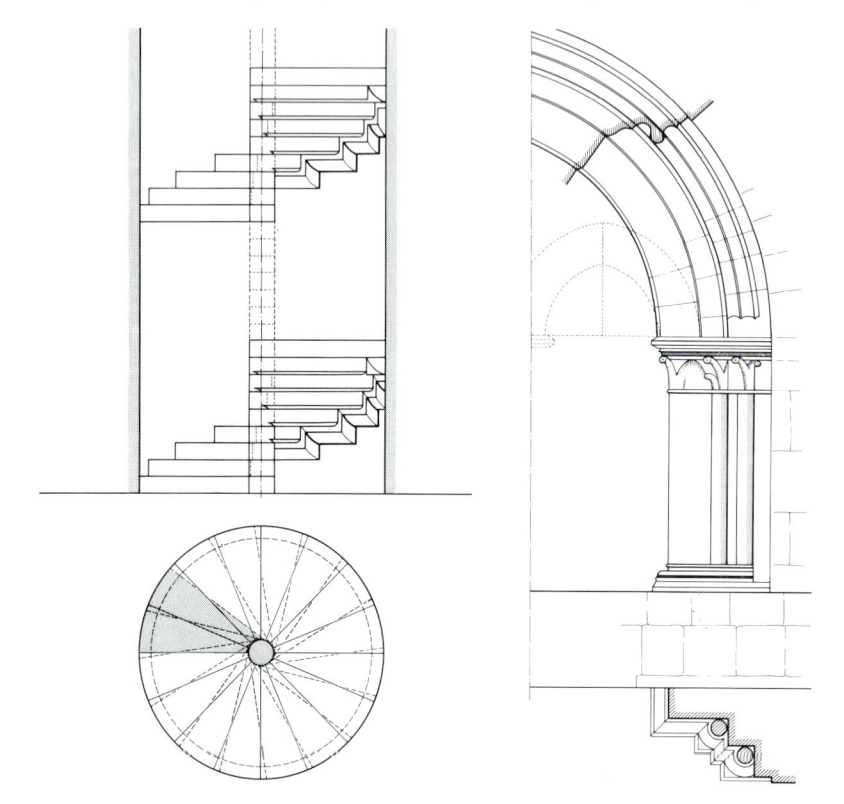

34   Syracuse, Sicily:
Vertical section through the staircase in the west tower of Castel Maniace.

35   Syracuse, Sicily:
Detail of window on the southwest side of Castel Maniace.

38   *Lines of Development of Hohenstaufen Architecture*

36 Syracuse, Sicily:
Interior view of Castel Maniace;
entry wall on the northwest side.

37 Syracuse, Sicily:
Southeast side with east corner tower
of Castel Maniace. Striking are the regular
cut of the squared stones and the plastic
concept of the slanted openings of the
narrow windows.

Castel Maniace

38 Syracuse, Sicily: Castel Maniace. Capital above one of the bundled columns in the open middle square.

39 Female head (detail of Fig. 38). The work shows a master's hand.

40 Syracuse, Sicily: Castel Maniace. Capital above the second restored bundled column in the open middle square.

41 Syracuse, Sicily: Capital of a column of the restored vault.

40   *Lines of Development of Hohenstaufen Architecture*

42 Syracuse, Sicily: Castel Maniace. View through the two restored rows of vaults. In the middle, right, two of the four bundled columns that were arranged around the open middle square.

43 and 44 Syracuse, Sicily:
Castel Maniace. Two of the four bundles, consisting of three slender columns each, that stood at the corners of the middle square of the hall.

45 Syracuse, Sicily:
Castel Maniace. Base of the column bundle shown in Fig. 43.

42   *Lines of Development of Hohenstaufen Architecture*

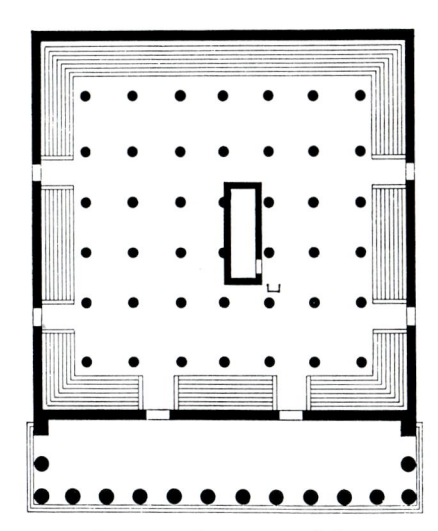

46 Eleusis: Telesterion of the temple of Demeter from the fifth century B.C. with the vestibule of Philon (drawing after G. Gruben).

of Castel Maniace; rather, they exhibit an artful composition of geometric forms that provide a much less tangible spatial experience.

The shape of the ground plan harks back to the fifth century B.C. and the *telesterion* of the temple to Demeter at Eleusis (Fig. 46). However, the origin of the square-type plan with the four corner towers is not clear; it has been discussed repeatedly. We will go into the question of origin in detail (p. 48 ff.), following a description of the other two citadel structures on the east coast of Sicily: Augusta and Catania (Castel Ursino).

With regard to the specific purpose of Castel Maniace, it has rightly been pointed out that it is not actually a military structure and therefore would be better named a palace, although the strategic significance of its location indicates that it was meant to be used for defense in case of war. In later centuries fortification-like outworks were added to the citadel for this purpose.

3 Augusta

Castel Augusta,[35] one of Frederick's works, situated between Castel Ursino and Castel Maniace, bears no resemblance today to its original form either outside or in the interior; however, it can be reconstructed with some degree of certainty. Judging by its dimensions, it was the largest of the edifices erected by Frederick II, with sides approximately 62 m in length, and probably one of the most magnificent. It was most likely begun in 1232, and there are indications that it stands on the walls of a Norman or even earlier fortress. Reconstruction attempts show an almost square plan including not cylindrical, but rectangular towers that do not lie strictly on the diagonal of the main rectangle (Fig. 47): they are shifted somewhat toward the middle of the north-south axis, so that they project less conspicuously from the walls on the east and west sides than they do with respect to the north and south faces. Accordingly, flat, rectangular towers are provided on the east and west wings, while on the north and south sides

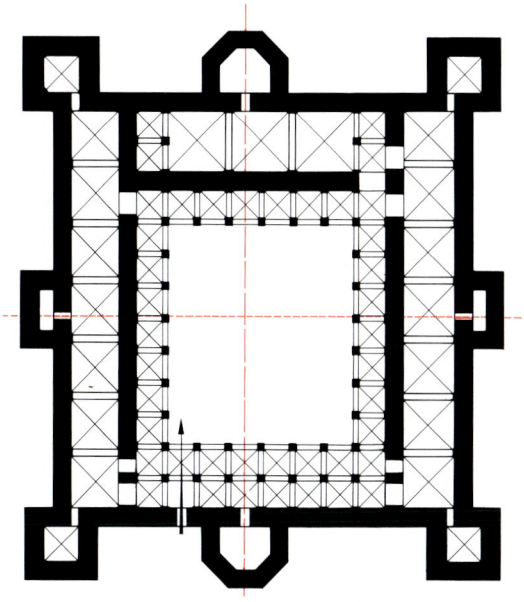

47 Augusta, Sicily: Ground plan of the citadel with lines of symmetry.

43

truncated octagons are positioned midway; these take up the rhythm of the more strongly protruding rectangular towers to the north and south through their own accentuated projection. If the lines of the square corner towers are included in the plan, the result is a rectangle flattened from east to west. In this way, the manifold symmetrical relationships that can be observed in Syracuse and Catania are reduced to a bilateral symmetry along the main north-south axis and along a secondary east-west axis. Along the diagonals the symmetrical relationship is reversed.

The deciding conceptional difference between the plan of this citadel and those of Castel Maniace and Castel Ursino lies in the basic form that is not strictly square, the rectangular corner towers, and the rectangular and hexagonal side towers. These features distinguish the layout of this stronghold from that of the other two fortresses, whose rigidly square form with round corner towers we will trace back in the following chapter to the tradition of the desert palaces and *ribāṭāt* of the Omayyad dynasty.

At Augusta another Byzantine tradition, likewise derived from the Roman-type *castrum*, was apparently followed, as A. M. Schmidt has pointed out.[36] It is possible that the citadel of Augusta goes back to forerunners of the Byzantine period (see also p. 57, Fig. 69).[37]

4 Castel Ursino

The best preserved of the three large citadels, still viewable today, is Castel Ursino in Catania.[38] No one remains unimpressed by the massive, plastically articulated structure (Fig. 49), whose cylindrical corner towers and semicylindrical side towers accentuate the again rigidly square basic figure.[39]

The plan exhibits a remarkable regularity (Fig. 48). The tendency to arrange the building components symmetrically and the translation of heavy building blocks into stereometric figures are even more pronounced here

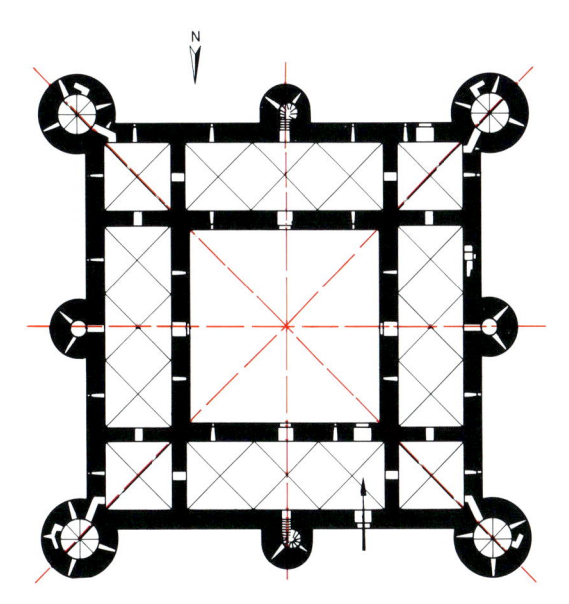

48 Catania, Sicily: Ground plan of Castel Ursino, with lines of symmetry in red.

49 Catania, Sicily: General view of the north front of Castel Ursino.

than at Castel Maniace. Over and above the square plan and the rigid correspondence of four symmetrical axes, homeomorphic symmetrical relationships appear, inasmuch as the form of the building as a whole is repeated in the square of the inner courtyard and is refracted again in the four square corner rooms. This system of homeomorphic symmetries appears here for the first time in a decisive form. We will find it again at Castel del Monte. If we consider the square of the inner court of Castel Ursino in connection with the four squares of the corner rooms, we obtain a simple, four-sided system of the sort that appears in far more complicated, and octagonal, purer form in Castel del Monte. The network of ribbed groined vaults spread over the entire structure continues harmoniously from one interior area to the next. The corners of the rectangular layout are accented by cylindrical towers that, with their pure circular shape, form a taut contrast to the basically square system of the citadel itself. The additional cylindrical side towers emphasize this effect pointedly.

This shows, in turn, how clearly these citadels evolved from the *ground plan*. It confirms again the observation that they were not adapted to the geological circumstances of the given location, but followed exclusively the idea that preceded the building and its mathe-

46   *Lines of Development of Hohenstaufen Architecture*

50   Catania, Sicly:
View of the northwest
corner tower of Castel Ursino.

51   Catania, Sicily:
West wing of Castel Ursino.

matical-geometric structural skeleton. It is logical that the geometry of the foundation is carried over into the stereometry of the building itself. The special power of the basic design gives these buildings an extraordinarily impressive plastic effect. In contrast to Castel Maniace, this citadel, begun after 1239, was built with irregular rubble stones — presumably because money was scarce.[40]

The external appearance of Castel Ursino is defined by the dominating form of the cylindrical corner towers (Fig. 50), which stand — similar to those in Syracuse — on conical bases. The motif of these towers is repeated in the smaller two-third cylinders on the wings and dies away in them (Fig. 51). The entry portal cannot play the same dominant role in this arrangement of towers as it does at Castel Maniace; it lies outside the symmetrical system, more or less eccentric, in the western part of the north facade between a corner tower and a side tower — as in the citadel of Augusta and later in Prato.

Castel Ursino   47

The general layout of the citadel is arranged along the axes formed by the four directions of the compass. Its mighty walls have huge unadorned areas that increase the impression of solidity, which is not lessened by two-dimensional decorative elements. We will return later to this particular feature of the buildings erected by Frederick II.

The newels in the cylindrical corner towers are shaped as *eight-sided* hollow prisms (Fig. 52),[41] encompassing 16 steps like those of Castel Maniace and Castel del Monte.

In 1948/50, W. Krönig was the first to call attention to the possibility that this Hohenstaufen type of four-wing layout originated in the region of Syria.[42] The late Roman defensive construction, the *limes arabicus*, under the emperor Diocletian in particular, was actively continued during the early Islamic period in Syria under the Omayyads, albeit with the addition of Syrian and Sassanian building concepts. Krönig and later authors answer the question of how the style found its way from Syria to Sicily by suggesting that Frederick II could have become familiar with such defensive structures during his crusade in the Orient. At the same time, there are some doubts about whether the emperor actually saw the Syrian structures — no information to this effect has been handed down. C.A. Willemsen therefore questioned this theory and, after pointing briefly to existing pre-Hohenstaufen four-wing constructions in Apulia, for instance in Bari — which basically would mean only moving the question forward in time, but not answering it — he came out decidedly in favor of a northern origin for this type of construction.[43]

Willemsen offers not very persuasive examples from Great Britain and northern France which he describes as "at least rudimentarily" regular four-wing constructions; however, they are simple walled strongholds that can be traced back to the Roman *castrum* in the region of England as well.[44]

Willemsen accounts for the lack of convincing connecting links in central Europe between British fortresses and the Hohenstaufen citadels in Italy, to which he himself calls attention, by the fact that the central European fortresses, especially those in Germany, built on mountainous heights, had to be adapted to the landscape. To counter this, it may be said that Castel Rocca di Calascio, 1460 m high in the Abruzzi, southeast of L'Aquila in

52 Catania, Sicily: Octagonal vault over spiral staircase in the southwest tower of Castel Ursino, with newels designed as eight-sided hollow prisms encompassing 16 steps.

48   *Lines of Development of Hohenstaufen Architecture*

the Hohenstaufen area, lies by no means on suitably even ground; rather, it was forced upon a sharp-edged rocky ridge (see p. 79, Fig. 111). Thus the different ground plans *cannot* be explained by the shape of the terrain — Castel Maniace does not conform to the land either; they result from structurally different concepts of design.

We return then to W. Krönig's reference to Syrian buildings as the prototypes of Hohenstaufen four-wing structures. What was the road that led from Syria to Sicily? Before answering this question we should change our angle of observation for the moment and consider Sicily not only from the north, but also in its position at the center of the Mediterranean basin. During those centuries of the Middle Ages the Mediterranean was bounded on *three* sides, on the east, on the south, and on the west, by the dominions of Arab dynasties (see map, Fig. 53). The southern tip of Sicily is only about 150 km away from the northern coast of Africa! For this reason very close ties have existed between North Africa and Sicily since time immemorial. The Phoenicians settled there and founded Palermo in the seventh century B.C. It was not until 254 B.C. that the Romans took it away from them. The alter-

53 Map of the Mediterranean.
The red hatching along the eastern, southern, and western coastlines shows the extensive area of Arab dominion in the middle of the thirteenth century. Place names in italics mark desert castles, citadels, and *ribāṭāt* of Omayyad origin, preserved up to the present, that are mentioned in the text.

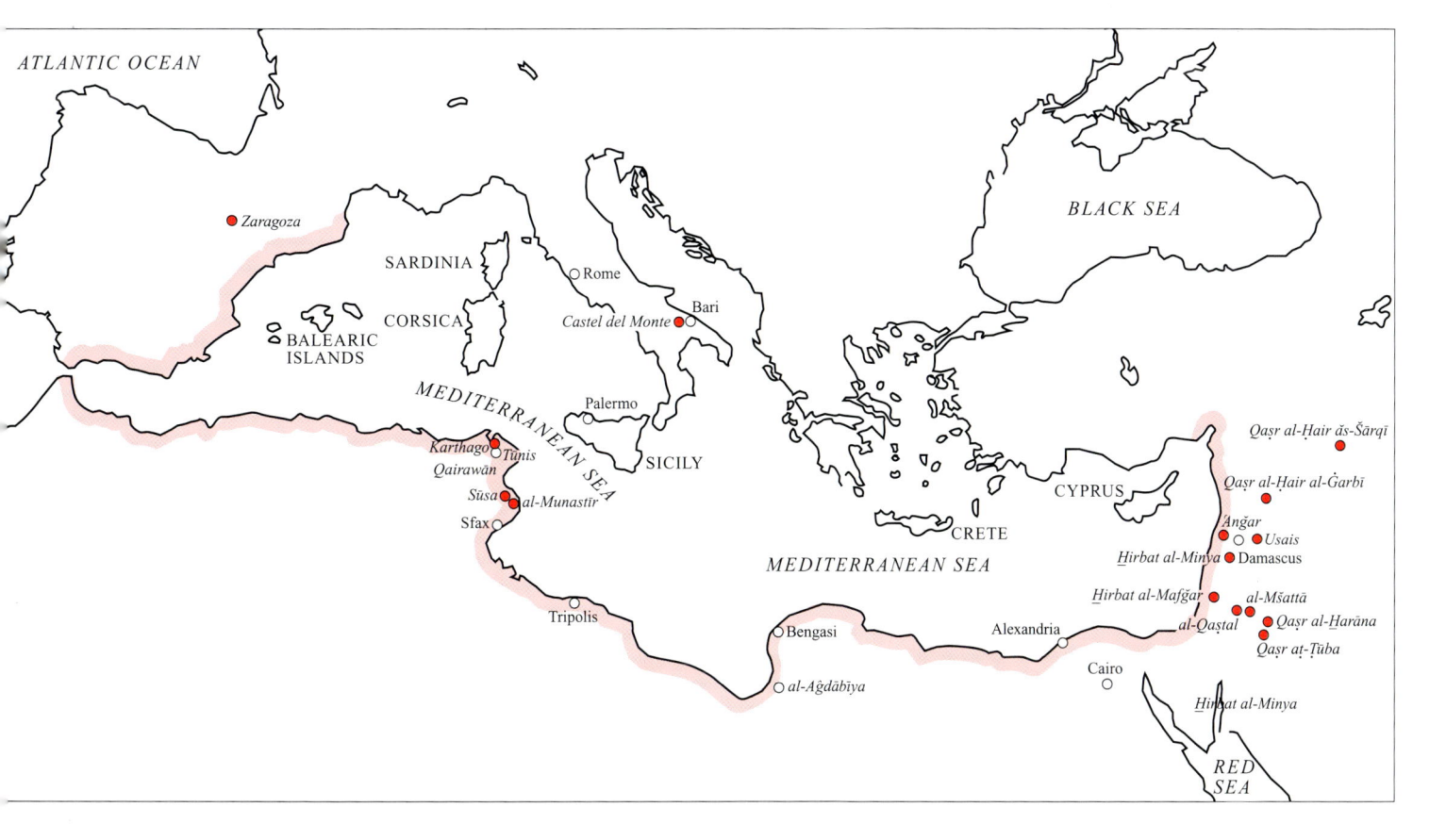

Castel Ursino 49

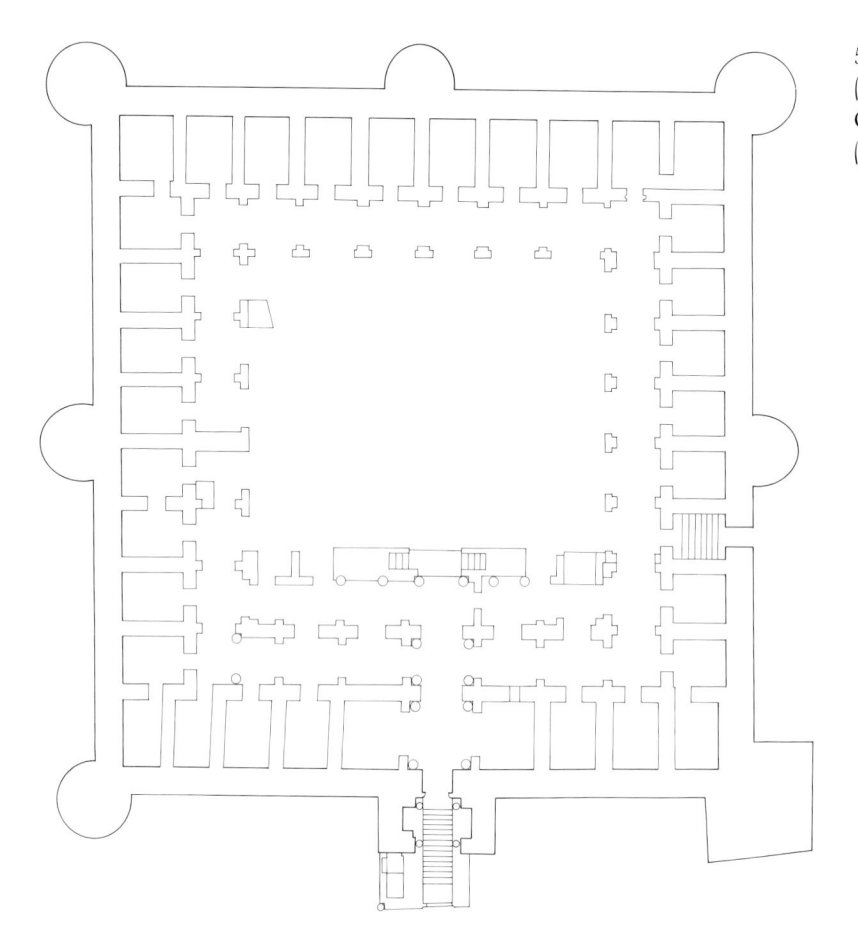

54 *Ribāṭ* at Sūsa (earlier than 796?), Tunisia: Ground plan (from Lézine 1956, Fig. 14).

nately friendly or antagonistic relations between Sicily and North Africa remained lively. Roger I concluded a treaty of friendship with the ruler of Tunis in 1075. From his base in Sicily, Roger II (1101–1154) controlled the coast from Tripolis to Cap Bon and land inwards to Qairawān![45]

Approximately 300 years earlier, an insurgent Byzantine admiral requested the help of the Aġlabides of Tunis in opposing Byzantine rule, long after trade relations had been established between Sicily and North Africa. In answer, the fleet of the Aġlabiden put out to sea from Sūsa (Sousse) on June 15, 827 A.D. to conquer Sicily and landed three days later, on June 18, at Mazara on the island's southwestern point.[46]

Some 150 years thereafter, the Arab geographer Ibn Ḥauqal,[47] who was in Sicily in 973 A.D., reported the following in his description of the earth, 'Ṣūrat al-arḏ' (The Shape of the Earth): "On trouve dans l'île [Sicily], au bord de la mer, beaucoup de couvents militaires: on y practique [sic] faussetè et hypocrisie...."[48]

50   *Lines of Development of Hohenstaufen Architecture*

55 *Ribāṭ* at Sūsa (earlier than 796?), Tunisia: the tilt of the minaret is an optical effect of the necessarily short distance at which the picture was taken (source: Edition Kahia, Tunis).

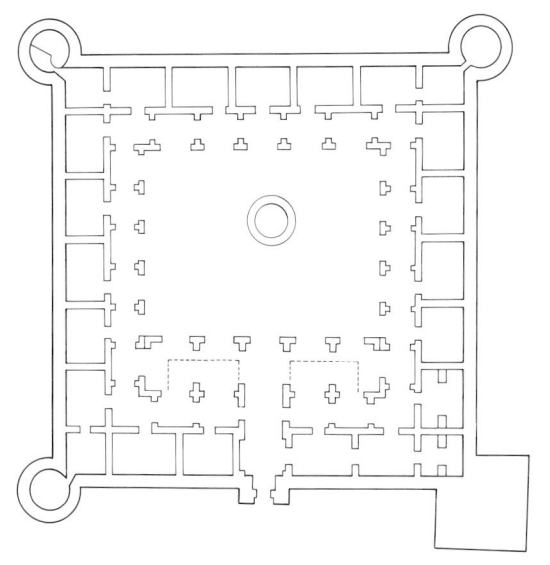

56 *Ribāṭ* at al-Munastīr (796 A.D.), Tunisia: Ground plan (from Lézine 1956, Fig. XXXIII b).

The word translated as "couvents militaires" was in the original *ribāṭāt*, the plural of *ribāṭ*. This unequivocal and reliable source takes us a great step forward, because there is no reason not to believe that these *ribāṭāt* looked exactly like those of Sūsa, al-Munastīr (Figs. 54–56), and Lamṭa, in the region from which the conquerors came. At the same time, this would mean that Arab *ribāṭāt* were built in Sicily long *before* the period of Hohenstaufen rule. It would also explain the fact that pre-Hohenstaufen buildings in Lower Italy had already been designed with four-wing layouts (for instance, in Bari,[49] as mentioned above). Frederick II frequently erected citadels on top of existing older structures (e.g., Oria; p. 29, Fig. 14). Only excavations can answer whether this was also the case in Syracuse or Catania.

The typological derivation described does not mean that, both in its general structure and in the individual forms of the citadels visible today, Hohenstaufen architecture did not develop its own style. It must merely be acknowledged that the Arabic component common to the architecture of the Norman period continued to operate to a certain extent during the Hohenstaufen period. In view of the numerous Arab-Norman buildings that were surely still in good condition at that time and Frederick's favorable attitude toward Islam, this would be natural. The widespread occurrence of Islamic ornamental art beyond the boundaries of the Islamic world bears this out (see p. 91, Figs. 129 and 130).

What is the building tradition behind the Tunisian *ribāṭāt* that were erected during the eighth and ninth centuries A.D., however? Research in the field of Orientalism

Castel Ursino 51

has already shown that the Tunisian *ribāṭāt* represent a type that developed within the sphere of influence of the dynasty of the Omayyad caliphs. During the lifetime of the prophet Muḥammad, the Omayyads were still one of the leading families of Mecca. In the year 661 A.D. Muʿāwiya I made his appearance as the first caliph in Damascus. During the period that followed, the Omayyads erected numerous castles and palaces in the Syrian desert, some of which can still be seen today (Figs. 57–66).[50] The militarily secured territory of the Omayyads at that time stretched from the Amur River to southern France. After the dynasty was expelled from Damascus by the Abbasids of Baġdād in 750, ʿAbd ar-Raḥmān I escaped the bloody persecution and, from 756 on, continued the Omayyad rule as emir of Córdoba. This emirate was elevated to a caliphate in 929.

The path taken in the expansion of power from Syria over Egypt and North Africa to Spain is studded with structures of a uniform basic type which — as already mentioned — had taken over the fundamentally rectangular form from the *castra* of the Roman *limes arabicus* and provided it with round corner and side towers. The plan of the Omayyad city of ʿAnǧar (714–715 A.D.; Fig. 57) clearly shows the arrangement of the Roman *castrum* with *cardo* and *decumanus*. The interior design of the desert castle al-Ḫarāna (710 A.D.; Figs. 59 and 60) shows Sassanian features. The palace of the caliph al-Walīd I in Usais, some 110 km east of Damascus (705–715 A.D.; Fig. 58), demonstrates a 'classically regular' symmetrical division of the interior.

A number of the Omayyad structures that have survived over the centuries are shown here on a map (Fig. 53) which at the same time indicates Sicily's geographically central position in the Mediterranean basin, directly opposite Islamic North Africa. The *ribāṭāt* at Sūsa, al-Munastīr, and Lamṭa are part of the chain described, which is in direct succession to the Syrian buildings. Originally, the number of structures representing the power and expansional force of the Omayyads was surely much greater. The chain continued westward of Tunis as far as Spain, where a late descendent from the eleventh century fortunately survives, in the form of the castle La Aljafería in Zaragoza (Fig. 67). The interior cupola in the shape of an eight-pointed star, similar to the one at Córdoba (p. 126,

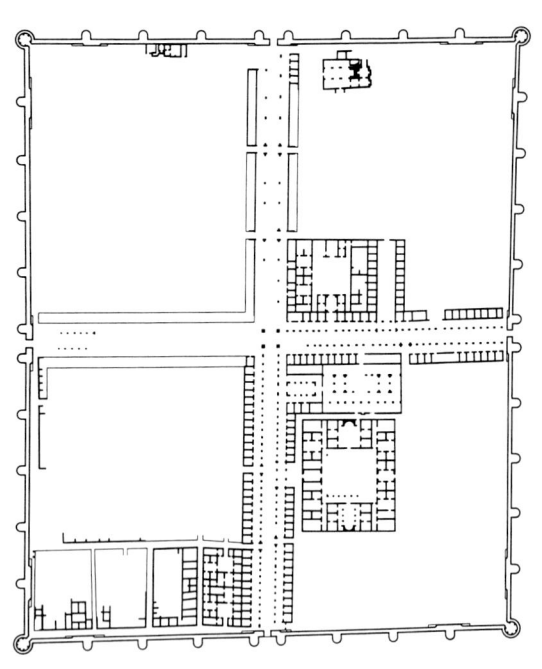

57 Omayyad city of ʿAnǧar (714–715 A.D.), Syria: Ground plan (from Stierlin 1979).

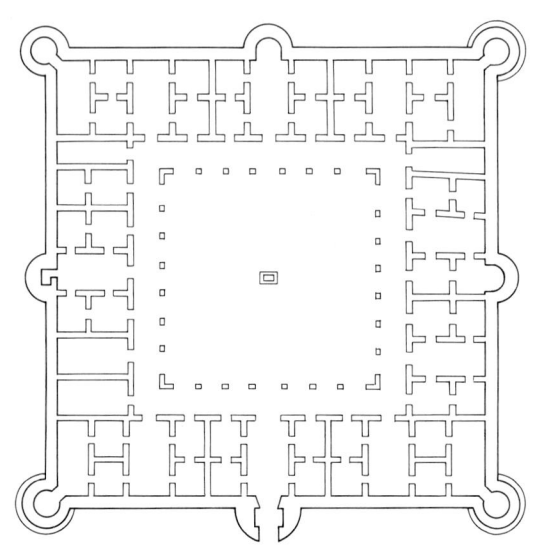

58 Usais, approximately 110 km east of Damascus, Syria: Palace of the caliph al-Walīd I (ca. 705–715 A.D.). Ground plan of the square building with sides of approximately 67 m (after *Propyläen Kunstgeschichte*, vol. 4, p. 155, Fig. 13).

52   *Lines of Development of Hohenstaufen Architecture*

59 Qaṣr al-Ḥarāna (710 A.D.), Jordan
(from Stierlin 1979, p. 52, Fig. 17).

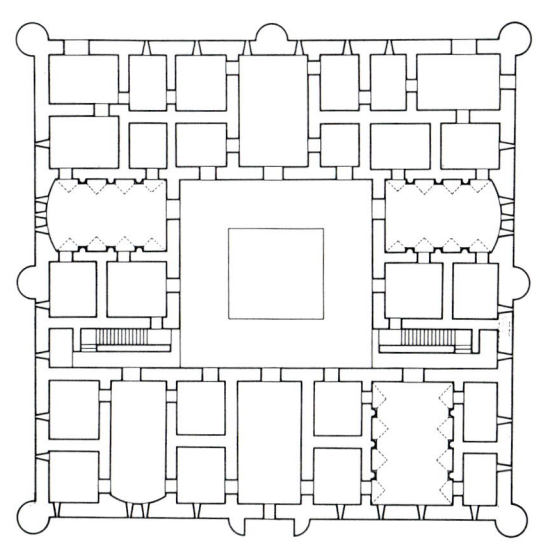

60 Qaṣr al-Ḥarāna (710 A.D.), Jordan:
Ground plan (from Stierlin 1979, p. 55).

Figs. 178 and 179), bears witness to Omayyad architectural style (see p. 127, Fig. 180).

With this chain of fundamentally uniform rectangular structures with round corner and side towers, reaching around the Mediterranean as it were, the pathway that this architectural style followed from Syria to Sicily also becomes clear. Moreover, it is evident that this typology was handed down long before the period of Hohenstaufen rule. Ibn Ḥauqal's written report confirms this. Important archaeological support for this transmission can be found in the evidence provided by A. M. Schmidt of an Arabic four-wing construction with round corner towers in western Sicily.[51] On the basis of accounts by the Arab geographer Ibn Idrīs at the court of King Roger II, and following careful studies and her own observations, Schmidt discovered the ruins of an undoubtedly Arabic citadel in Mazzallaccar near Sambuca that is enclosed by sturdy walls, 1.10 m thick, and has four round corner towers topped with cupolas (Fig. 68). Both the height of the walls and towers and the tower diameters are 5 m. The almost square building measures 51.60 × 54.20 m. In connection with this discovery, Krönig has rightly pointed out how sketchy is our knowledge of Sicily's artistic culture during the Arab period that lasted more than two centuries. How much has been lost can be inferred from the manifestly quick decay of

Castel Ursino 53

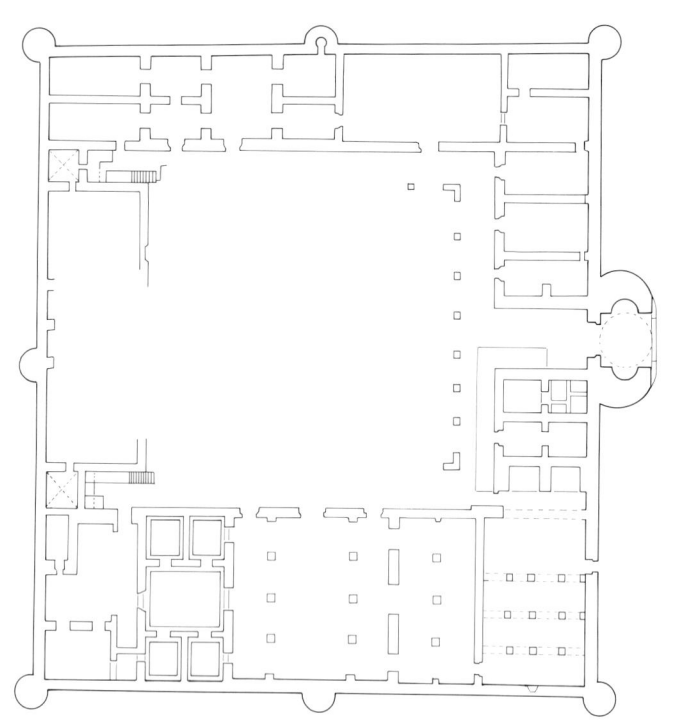

61 Ḥirbat al-Minya (ca. 705–715 A.D.), Syria: Ground plan (from *Propyläen Kunstgeschichte*, vol. 4, p. 156, Fig. 15).

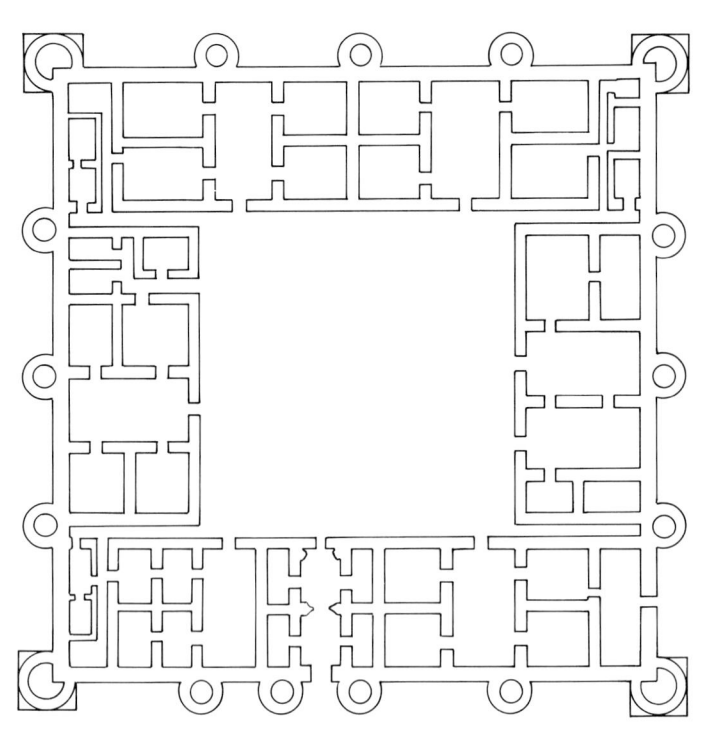

62 Al-Qaṣtal (after 720 A.D.), Jordan: Ground plan (after Stern 1946, Fig. 10).

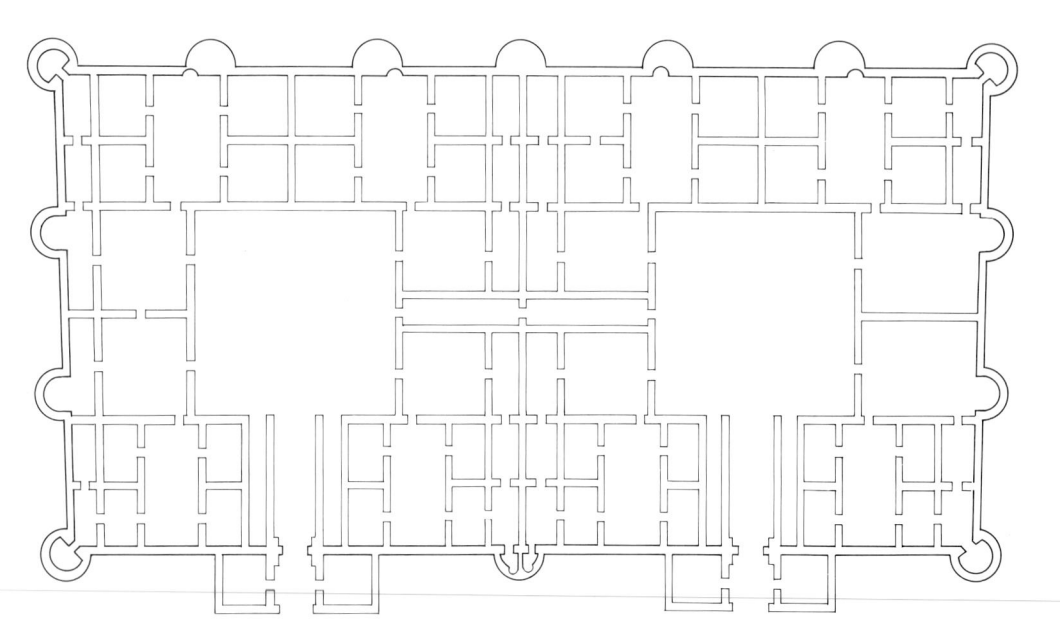

63 Qaṣr aṭ-Ṭūba (743–744 A.D.), Jordan: Ground plan (after Stern 1946, vol. 11–12, Fig. 5).

54   *Lines of Development of Hohenstaufen Architecture*

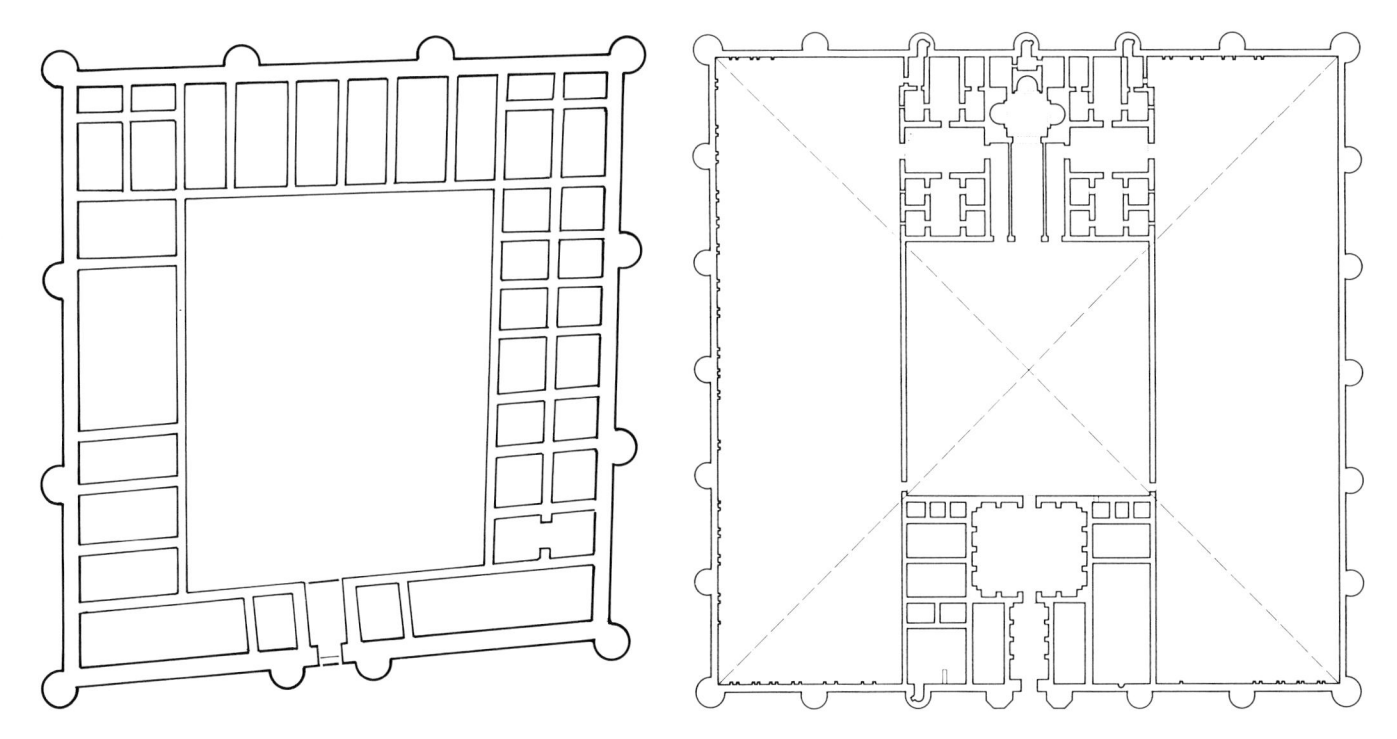

64 Qaṣr al-Ḥair aš-Šārqī (first quarter of the eighth century), Syria: Ground plan.

65 Omayyad palace Mšattā (before 750 A.D.), Jordan: Ground plan (from Stierlin 1979, p. 54).

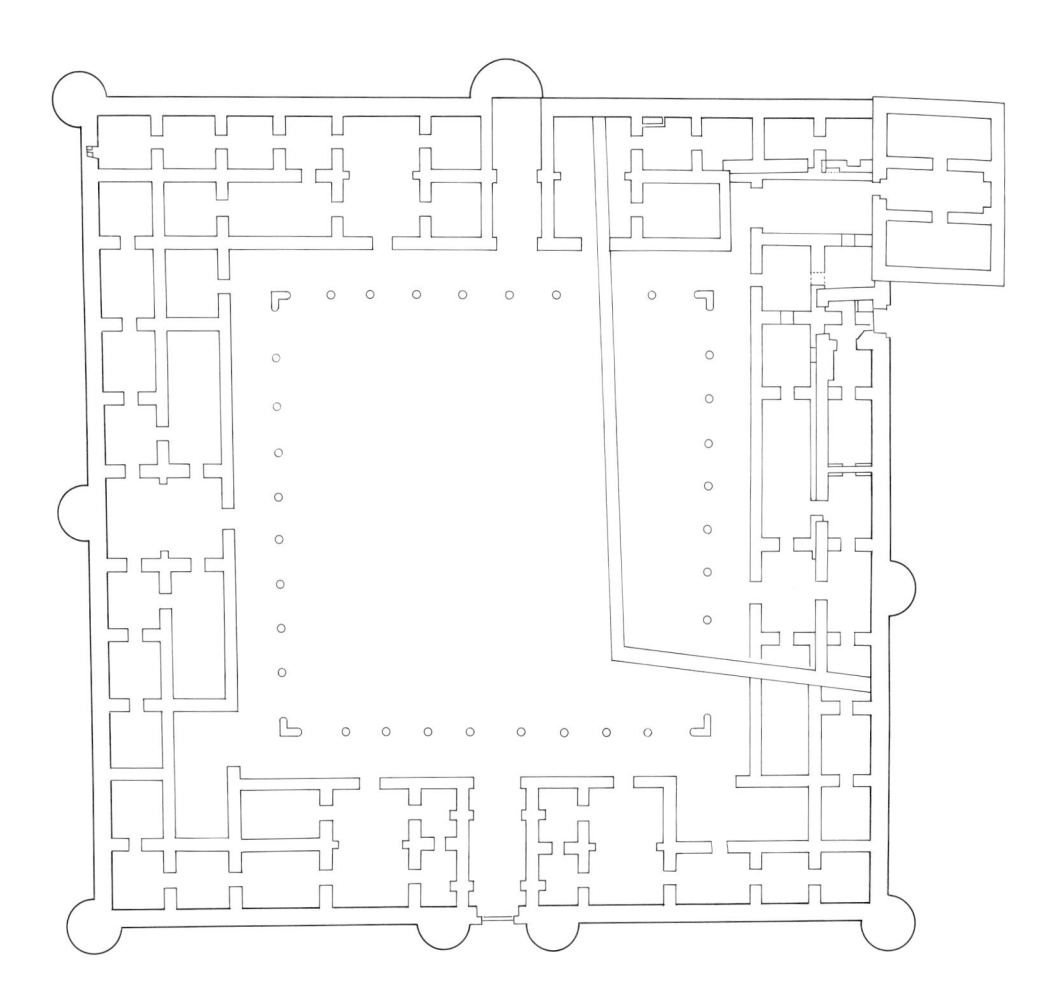

66 Qaṣr al-Ḥair al-Ġarbī (second quarter of the eighth century), Syria: Ground plan (after *Propyläen Kunstgeschichte*, vol. 4, p. 170). See also the portal, Fig. 95.

Castel Ursino 55

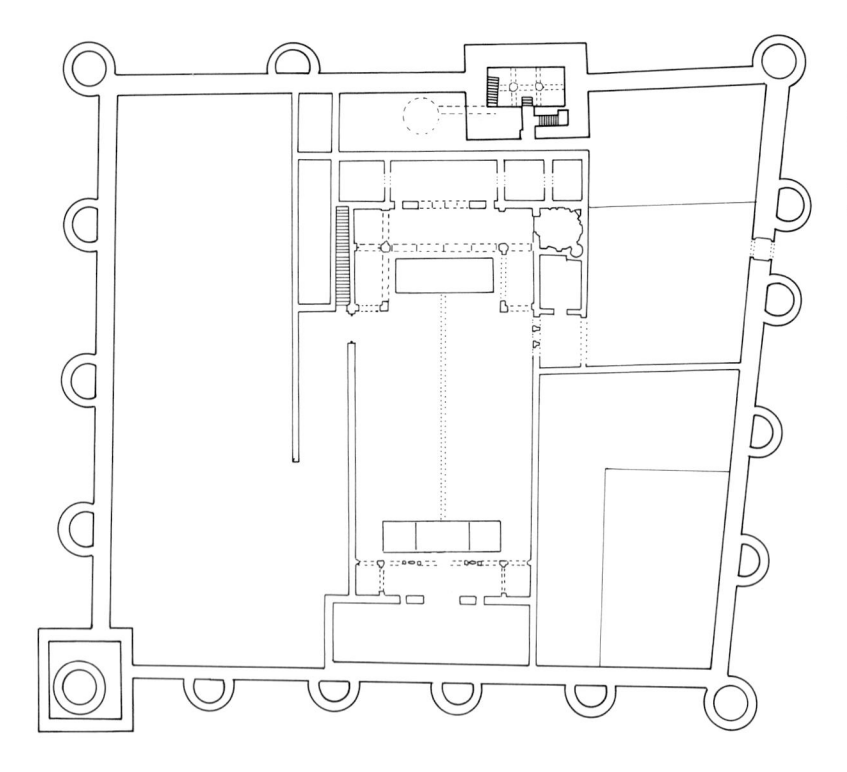

67 Al-Ġaʿfarīya, Zaragoza, Spain (La Aljafería, second half of the eleventh century): Ground plan (after *Propyläen Kunstgeschichte*, vol. 4, p. 225, Fig. 49).

the old remains — consider Fig. 11 in the publication by A. M. Schmidt, showing a still-standing minaret; it collapsed in 1968 and since then has been going completely to ruin. It is not difficult to imagine that Frederick II himself saw numerous interesting buildings from the time of the Arabs in Sicily that were still standing erect.

A new discovery was made recently that is of similar importance for the pre-Hohenstaufen architecture of Sicily: Between the Greek temples A and O at Selinunt, D. Mertens has identified a medieval 'fort',[52] the form of which is shown here in Fig. 69. With its rectangular but not square plan and its angular towers it is not related directly to the 'Omayyad' type represented by the *ribāṭ* at Sūsa, but rather to the so-called Byzantine type of fortified sites, of which C. A. Willemsen has compiled several examples.[53]

Thus, Selinunt is more closely connected with the Hohenstaufen Castel Augusta (p. 43, Fig. 47) than with Castel Maniace or Castel Ursino. However, the discoverer correctly calls attention to North African fortresses of the Byzantine period. Perhaps it would even be wise to consider the emergence of the 'fort' in Sicily's Byzantine era; Syracuse was once, though only briefly (660–665 A.D.), the capital of the eastern Roman empire![54]

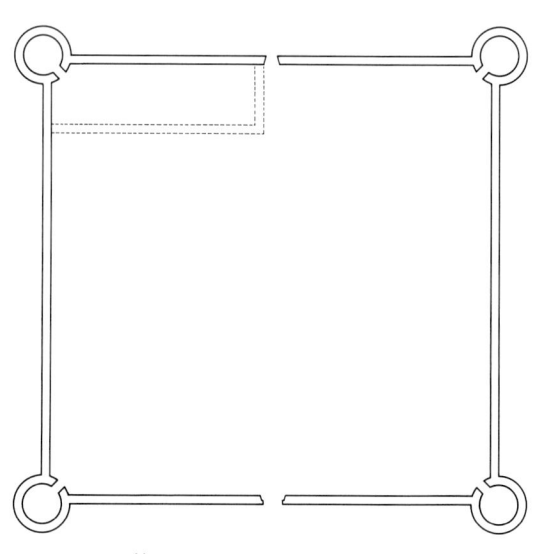

68 Mazzallaccar near Sambuca, Sicily: Ground plan of the Arab fortress (from Schmidt 1972).

56 *Lines of Development of Hohenstaufen Architecture*

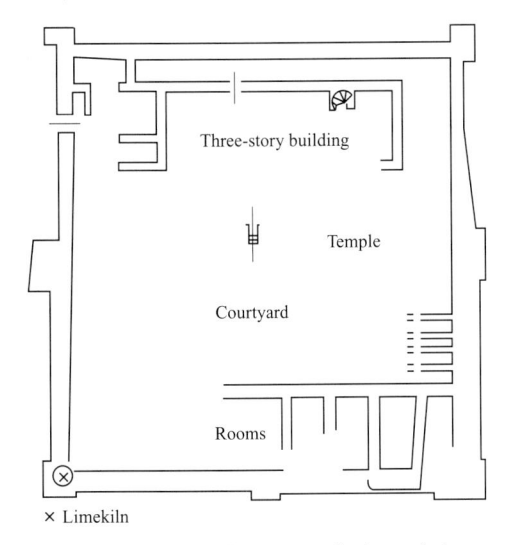

69 Selinunt, Sicily: Ground plan of the square fortress between the Greek temples A and O. Presumably Byzantine (after Mertens 1989, pp. 391–398, plate 37).

To summarize, it can be said that, together with Ibn Ḥauqal's written report, the history of architectural tradition described here permits us to conclude that the Hohenstaufen four-wing structures, as exemplified in Sicily by Castel Maniace and Castel Ursino, owed their form to Islamic prototypes, and that this Omayyad building type had already arrived in Sicily during the pre-Norman period. This explains the stylistic independence of Frederick's buildings, which do not directly follow the Omayyad tradition but are *indirectly* related to it via the Sicilian-Arab buildings, the key words being 'Ibn Ḥauqal' and 'Mazzallaccar'. The question of whether preserved or ruined *ribāṭāt* stood on the sites where the Hohenstaufen citadels were built must remain open or left to be answered by excavation research.

If we compare the plan of Castel Ursino (p. 45, Fig. 48) with that of the desert castle al-Ḥarāna (Fig. 60), we are surprised at the similarity between the outline of their walls, the arrangement of their towers, and the closely related basic layouts. At the same time, we note characteristic differences. Thus Castel Ursino is impressive because of the strong 'effect' produced by the masonry, which is quite different from that of the likewise, in its own way, magnificent character of the desert castle. The corner towers in Catania burst with the massive force of solidity, while the more slender towers of al-Ḥarāna are comparatively unobtrusive.[55] Al-Ḥarāna displays large, two-dimensional areas, Castel Ursino delimitations of stereometric elements that appear three-dimensional. Castel Ursino and Castel Maniace derive from the classical Greek tradition of three-dimensionality — inside as well as outside.

One feature of the *Islamic* building style was clearly passed on to *Hohenstaufen* architecture, however: the purity of geometric form, which also distinguishes it fundamentally from other European architecture; especially the fortresses of central Europe offer nothing to compare with this clear mathematical, geometric concept (cf. Château Chillon, Fig. 70). Characteristic features of Hohenstaufen architecture can be defined as follows: In its three-dimensional expression it follows the classical Greek, European tradition, thus contrasting with the Islamic. In its geometric structural concept, on the other hand, it is far removed from central Europe and closely

70 Château Chillon near Montreux in the canton of Waadt, Switzerland, situated on a cliff on the shore of Lake Geneva. Developed from a ninth-century fortress, its present form dates from the thirteenth century, when it served as the residence of the Count of Savoy. Its ground plan is characteristic for fortresses of that period in central Europe.

connected to the models of the Islamic sphere. We will examine in more detail the reasons for the basic mathematical relationships that reveal the connection to Arabic architecture when we consider Castel del Monte. However, an interesting and important observation should be called attention to here, concerning the two Sicilian citadels and Lucera. The system of dividing the interior areas of the Hohenstaufen buildings corresponds to a planning grid of 5×5 squares (Fig. 71). At Syracuse, the pattern covers the interior entirely; at Catania, an outer series of sixteen yoked squares surrounds the inner court, which takes up the area covered by the nine remaining squares. This pattern is repeated in the plan of the palace prism of Lucera, which follows the Pythagorean theorem: $a^2 + b^2 = c^2$ (Figs. 72–74). The cases in which this expression can be represented by natural whole numbers — for example, $3^2 + 4^2 = 5^2$ (9 + 16 = 25) — are infrequent. These 'Pythagorean triples' were just as familiar to the Arab mathematicians as they were to Leonardo of Pisa. We must work on the assumption that they were described in the Arabs' widespread

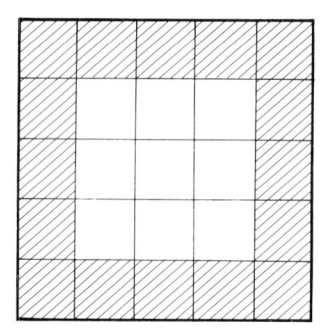

71 Schematic pattern of $5 \times 5 = 25$ squares. The outer, hatched row contains 16 squares, nine remain on the inside. This division plan corresponds to the Pythagorean theorem: $a^2 + b^2 = c^2$, or $3^2 + 4^2 = 5^2$. Integral sequences that conform to this Pythagorean theorem are known as Pythagorean triples.

72–74 Ground plans of the citadels of Syracuse (Castel Maniace), Catania (Castel Ursino), and Lucera.

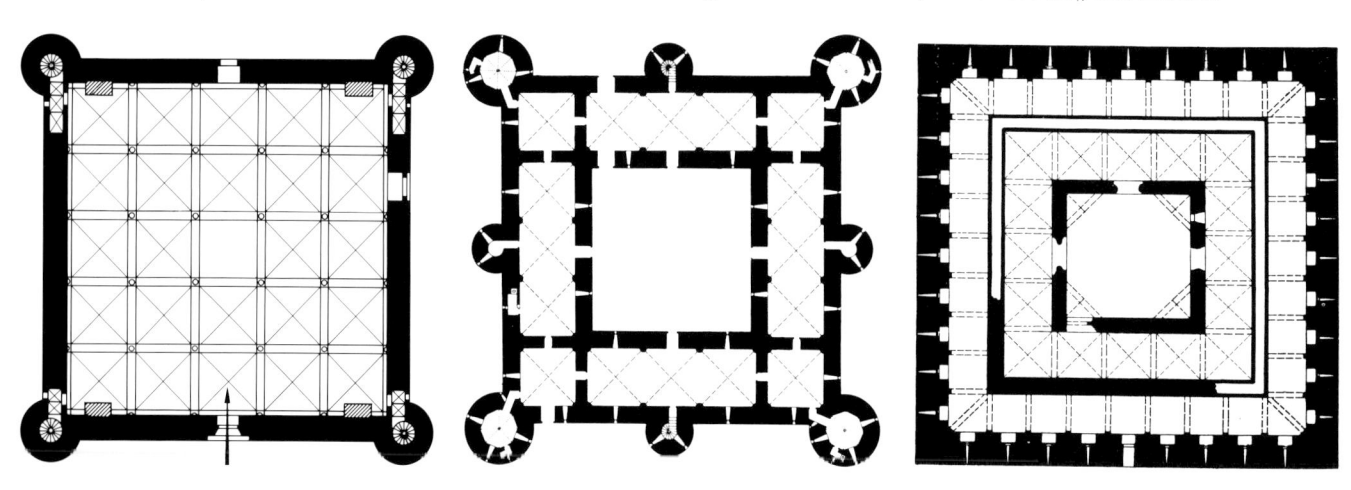

Lines of Development of Hohenstaufen Architecture

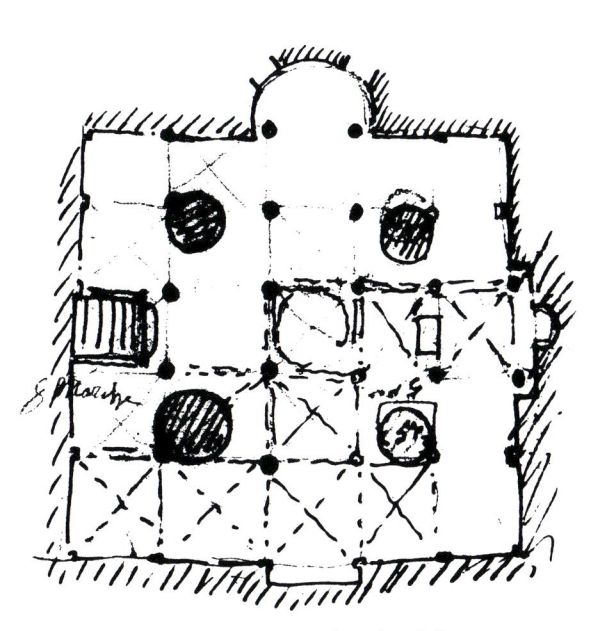

75 Louis Jean Desprez, sketch of the lower church of Santa Maria di Siponto (twelfth/thirteenth century; from Wollin 1935, p. 205, Fig. 45).

guidelines to practical geometry — especially as the Pythagorean triangle was already well known to the Babylonians and the Egyptians as an important surveyor's aid.[56]

The significance of this observation becomes clear when we realize how frequently this pattern occurs in other important buildings of the Arab-Islamic world and its sphere of influence. The crypt of Santa Maria Maggiore in Siponto, an old, formerly major port of the crusaders in Apulia and a center of trade with Egypt and Asia Minor, shows the same pattern of 5×5 squares. The French architect Louis Jean Desprez,[57] who made many drawings of architectural-historical interest, pictured in a sketch the pattern of the groined vaulting in the lower church of Santa Maria di Siponto (Fig. 75); according to the most recent observations, the vaulting was put in during a later building phase, presumably between the end of the twelfth and the beginning of the thirteenth century.[58] The structural pattern, which is different from that of the upper church, is artfully integrated into the system of massive substructures of the four pillars that carry the cupola.

The use of this pattern corresponding to a Pythagorean triple points to a knowledge of mathematics that can be traced to the Indo-Arab region, rather than to central Europe. Indeed, in the former area numerous other examples can be found. A contemporary structure is shown in Fig. 76: the plan of the Madrasa al-Firdaus in Aleppo, built in 1233.[59] The part of the building that is adjoined to the east strongly detracts from the regularity of the ideal solution for a foundation plan. Only eleven of the sixteen cupola-crowned squares that would be expected are therefore present. The symmetry of the courtyard, which repeated the 5×5 pattern in the ambulatory, is thus interrupted as well (see the broken lines in Fig. 76). In contrast, the Şaihzade mosque in Istanbul (Fig. 77), built by Sinān between 1544 and 1548, displays a very beautiful, regular solution of the 'Pythagorean triple' type. We see the same arrangement of the sixteen cupola-covered squares as in the schematic drawing (Fig. 71). This is by no means an isolated example. Classical mathematics was quite preoccupied with the diagonals of squares and rectangles, the irrational incommensurables. In his dialogue *Menon*, Plato dealt with the problem of duplicating squares, in which the diagonals play an important role. We find the duplication of the area of the square again in Vitruvius

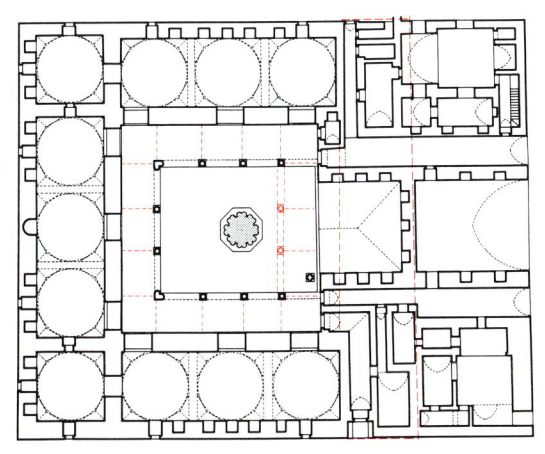

76 Aleppo, Syria: Madrasa al-Firdaus (from Stierlin 1979, p. 162).

Castel Ursino 59

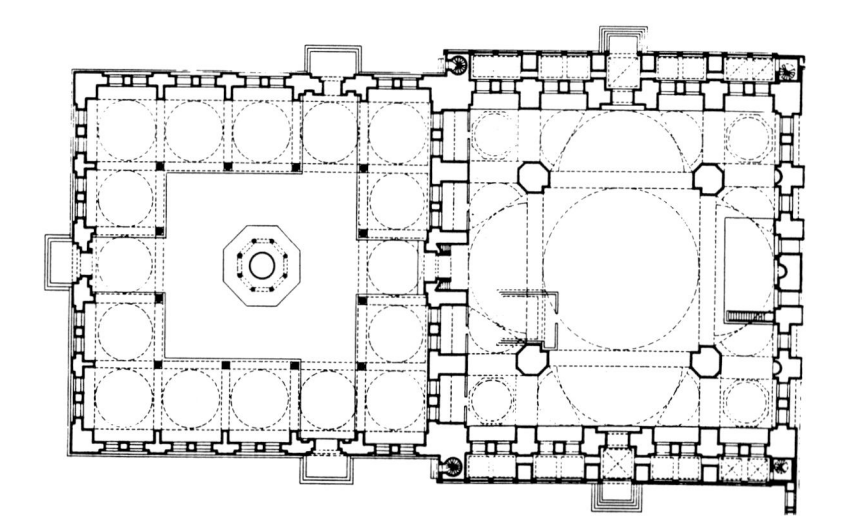

77 Istanbul: Ground plan of the Şaiḥzade mosque (1544–1548).

and in Master 2 of Villard de Honnecourt. Arabic decorative art and architecture — which cannot be separated from one another — were concerned again and again with this feature of irrationality, just as in their ornament the endless repetition of a pattern is chosen as a motif to symbolize infinity.

For our purposes, suffice it to say that the occurrence of the Pythagorean triple, with its complex relationships, in the ground-plan designs of Hohenstaufen buildings is an important indication of links to Islamic architecture, which are manifested not only in the exterior shape of the square citadel but also in the arrangement of the interior area.

5 Lucera

A very original architectural monument, belonging to the group of individual 'solitaires' of Hohenstaufen architecture, stood at Lucera,[60] not far from Foggia (Figs. 78–80), on the same mountain ridge that bore the settlement of the Saracens. Frederick II had transported them to this spot from Sicily following the Saracen war that began in 1222, in order to remove them from an area where they were continually the cause of disturbances. Frederick created a new homeland for them here. By means of administrative concessions and a policy of religious tolerance, and by

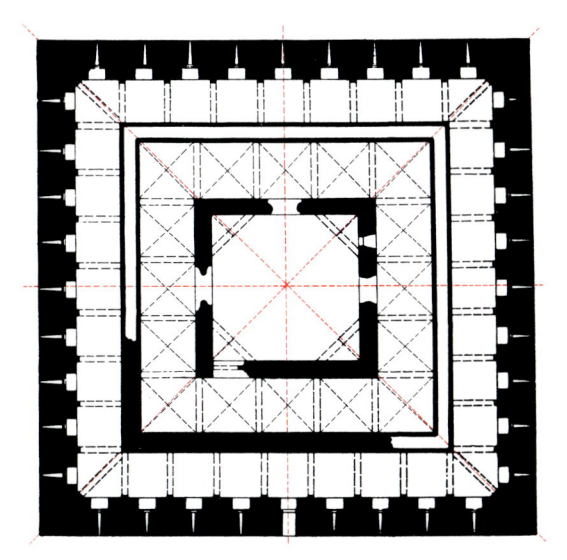

78 Lucera, Apulia: Plan of the citadel with lines of symmetry.

permitting them to further cultivate their manners and customs, he not only returned them to peace and order, but also secured for himself their devoted allegiance for all time. They showed the emperor their gratitude and furnished him from that time on with a loyal combat troop on which he was able to rely even in critical times. There are no reliable documents concerning when construction of the fortress-like castle began; it is set at about 1233. Unfortunately, the building is destroyed down to its foundations, but informative drawings by the French architect Louis Jean Desprez exist from the year 1778, i.e., prior to the blowing up of the ruins in 1790,[61] and the reconstructions by C. A. Willemsen are based on these.

The mountain ridge rising 250 m above the plane of Foggia virtually invited the building of a stronghold. Seen from the air, the citadel stands out clearly even today from the shape of the later Angevin fortress grounds (Fig. 80); it is marked, like the buildings previously considered, by a strictly square foundation. Above it rises a building that, in this form, exceeds every tradition. In it totally new, highly original building concepts were realized.

The square plan (Fig. 78) is familiar to us from Castel Maniace and Castel Ursino, as is the pronounced manifold symmetry. Here at Lucera everything appears to be striving toward a climax uniting traditional with new elements. Features of design become visible that disclose a further characteristic of Frederick's architectural style.

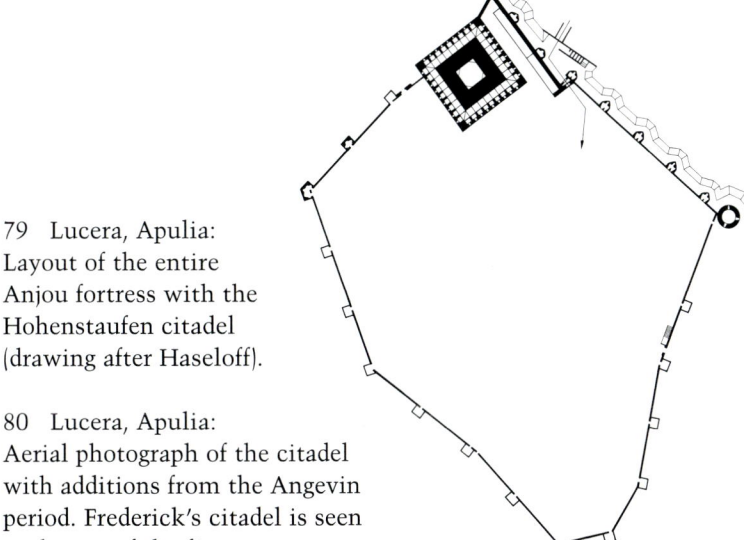

79 Lucera, Apulia: Layout of the entire Anjou fortress with the Hohenstaufen citadel (drawing after Haseloff).

80 Lucera, Apulia: Aerial photograph of the citadel with additions from the Angevin period. Frederick's citadel is seen at the top of the figure.

Lucera 61

The Norman-type residential tower can still be identified in the huge cubic prism that arises on the plateau of a truncated square pyramid. Here connecting lines can be traced from such structures as the Norman donjons in Adranò and Paternò at the foot of Mount Etna on Sicily (Figs. 81 and 82) that also lead to the citadel of Termoli, which will be discussed later (p. 77, Fig. 107).[62] Clearly visible here, as well, is the preference for stereometric forms with pleasure in the intersection and combination of different categories of solids.

Without the two very illuminating drawings by Louis Jean Desprez (Figs. 84 and 86) it would be impossible to imagine how the citadel of Lucera originally looked. The convincing reconstructions by C. A. Willemsen are based substantially on them (Figs. 83 and 85).

The expanded donjon is combined here with the pattern of the four-wing construction that has an inner courtyard in the shape of a square prism. In the uppermost of the three stories the square becomes an octagon (Figs. 83 and 85) — analogous to the Arabic architectural idea of changing the square of the ground plan via an octagon into a circular drum on which the cupola rests. There were also examples of this in the Arab-Norman churches

81 Adranò near Catania, Sicily: Norman residential tower (eleventh and twelfth centuries); cf. also Fig. 112.

82 Paternò near Catania, at the foot of Mount Etna, Sicily: Norman residential tower (eleventh and twelfth centuries).

62   *Lines of Development of Hohenstaufen Architecture*

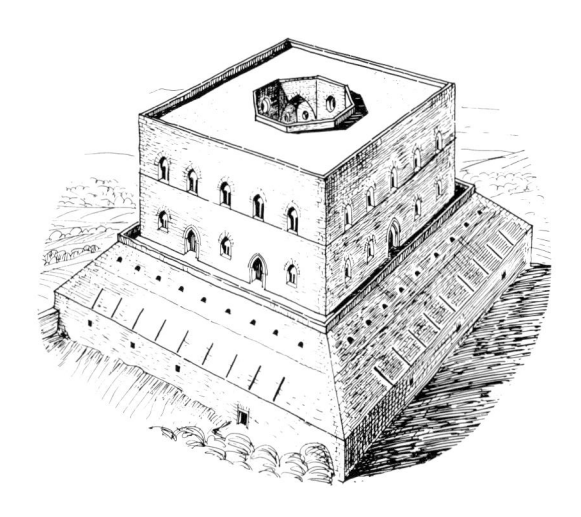

83   Lucera, Apulia: Reconstruction of the Hohenstaufen citadel (after Willemsen).

84   Louis Jean Desprez, drawing of the citadel of Lucera prior to its destruction around 1786. Total view (from Wollin 1935, p. 202, Fig. 38).

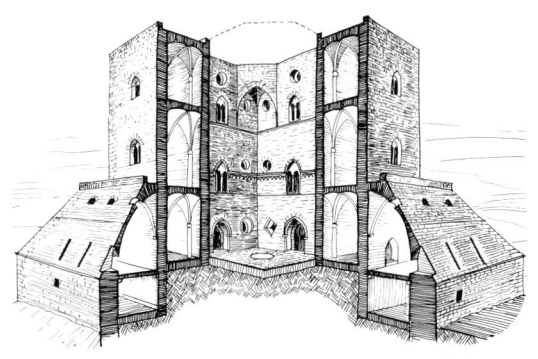

85   Lucera, Apulia: Section and view of the interior court (reconstruction after Willemsen).

86   Louis Jean Desprez, drawing of the remains of the interior court of Lucera still standing 1786 (from Wollin 1935, p. 203, Fig. 39).

of Palermo (cf. p. 23, Fig. 9). The novel realization of the uppermost story has been connected with the octagonal plan of Castel del Monte. The rich decoration of the citadel windows is striking and lends it the character of a palace more than of a fortress.

The drawings by Desprez show other important details as well, for example, the rhomboid ornamental recessed panels on the courtyard walls of the ground floor, alternating with round ones. This is a leitmotif that is found mainly in blind arches of Apulian, but also Pisan churches. Ernst Kühnel published a brilliant study on this subject, in which he points out how this motif connects the churches of Apulia and Pisa with the al-Ḥākim mosque in Cairo (Fig. 87). Desprez clearly accentuated these ornamental panels in his drawing. Perhaps they had already attracted his attention on other Apulian monuments that he had seen. In close vicinity are the cathedrals

Lucera   63

of Troia (Fig. 91) and Santa Maria di Siponto (Fig. 88), on which the motif also appears. Both churches date back to the twelfth century.[63] Desprez captured Santa Maria di Siponto in several drawings (Figs. 75 and 89). Finally, the church of Santa Maria Maggiore in Monte Sant'Angelo must be mentioned (Fig. 92).[64]

The same motif is found in Pisa and in the sphere of influence of Pisan art; it is at its most beautiful on the facade, sides, and transept of the cathedral (twelfth century, Fig. 90). In Pisa itself there are a number of other examples, such as San Nicola and San Paolo da Ripa d'Arno.

Kühnel pointed out that the motif appeared in Cairo a full century earlier than in Italy. There are still other, later examples of it in Cairo. Where it originated has not yet been determined. In Italy the occurrence in two self-contained regions that are geographically far removed from one another, in northern Apulia and in the area of Pisa, is remarkable. Both Pisa and Apulia traded commercially with Egypt. The examples in Apulia are more closely connected with those in Cairo and are also possibly somewhat older. In Cairo, as in Apulia, the squares standing on one of their corners are found in the lower part of the blind arches. Thus, Kühnel leaves undecided the question of whether the two regions, Pisa and Apulia, adopted the motif from Cairo independently of one another.

This motific connection, which has not been mentioned in the literature on Hohenstaufen architecture to date, appears as further evidence of the spread of Arabic-Saracen design in certain Italian regions. There are numerous other examples as well, such as on the cathedrals of Pisa and Troia (p. 91, Figs. 129 and 130). As in the Norman period, such motifs were most likely experienced as a lively enrichment of a people's own world of forms.

The mathematical division of the ground plan of the prism-shaped palace into a pattern of 25 squares — sixteen in the four-cornered building, nine in the courtyard — has already been discussed in connection with Castel Ursino and Castel Maniace. The strict geometric design of all three buildings becomes evident in this plan, as does the relationship to the mathematical-geometric thinking of Islam, which appears so forcible and at the same time artistically impressive in the geometric complexity of its architectural ornamentation. These mathe-

87 Facade of the al-Ḥākim mosque, Cairo (detail).

64 *Lines of Development of Hohenstaufen Architecture*

88 Santa Maria di Siponto, Apulia: North facade (present condition).

matical-geometric components that crop up in the outer form as well as in the design of the ground plan may be regarded as a particular characteristic of Hohenstaufen architecture.[65] It differs radically, on the other hand, in its uniquely plastic solidity and three-dimensional quality

89 Louis Jean Desprez, sketch of the north facade of Santa Maria di Siponto, Apulia (from Wollin 1935, p. 204, Fig. 42).

Lucera 65

from Arab-Islamic architecture, which tends to concentrate on surface as such.

The ground plan relationships we have established bring Lucera just as close to the Sicilian citadels as to Castel del Monte with regard to design history.[66] The castle Lucera, in which Frederick II was very fond of staying,

90 Pisa, Tuscany: General view of the cathedral (twelfth century).

91 Troia, Apulia:
North facade of the cathedral (twelfth/thirteenth century).

66 *Lines of Development of Hohenstaufen Architecture*

92 Monte Sant'Angelo, Apulia: Church of Santa Maria Maggiore, facade, presumably from the first half of the eleventh century, by Leone, Bishop of Siponto. The present condition probably dates to the end of the twelfth century (from Mola 1983, plate VI). Striking here also are the rhomboid ornamental panels!

must have been furnished with unusual splendor. Here the emperor's partiality for classical works of art, which he collected, was especially apparent. He gave orders to have them brought from Naples; they had to be carried over the mountains by strong men on their backs in order to reach their destination safely and undamaged.[67]

6 *The Gate to the Bridge of Capua*

A structure with quite another destination stands north of Naples: the gate to the bridge of Capua (Fig. 94).[68] The imperial secretary Riccardo da San Germano reports: "Imperator de Apulia venit in terram laboris et tunc ab ista parte Capue fieri super pontem castellum iubet, quod ipse manu propria consignavit." The concluding subordinate clause contains a highly significant statement regard-

67

ing the emperor's relationship to the structures he commissioned. Even if one can hardly infer from it — as has been done — that the design stemmed from the emperor's own hand, it follows with certainty that the blueprint personally *approved* by him was a realization of his own intentions and ideas.[69] It is permissible to deduce in general, from this account by Riccardo da San Germano, the interest and active participation of Frederick II in the designs of his architects.

Despite its ruined condition, even today the remains of this bridge citadel give the impression that was described in 1266 by Andreas of Hungary, the chronicler of Charles of Anjou, in these words: "two towers of astonishing power, beauty and size."[70] As with many of Frederick's buildings, it was not time alone but mostly deliberate defacement that caused its present state. Fortunately,

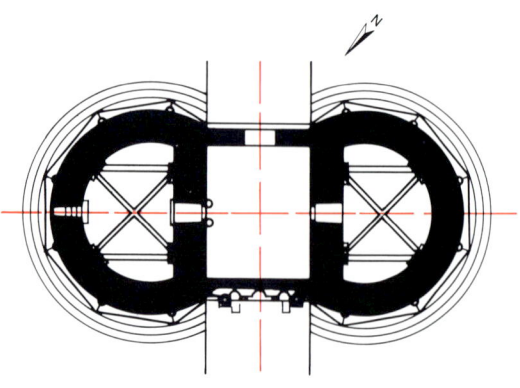

93 Capua, Campania: Plan of the upper story of the bridge citadel (from Shearer 1935), with lines of symmetry.

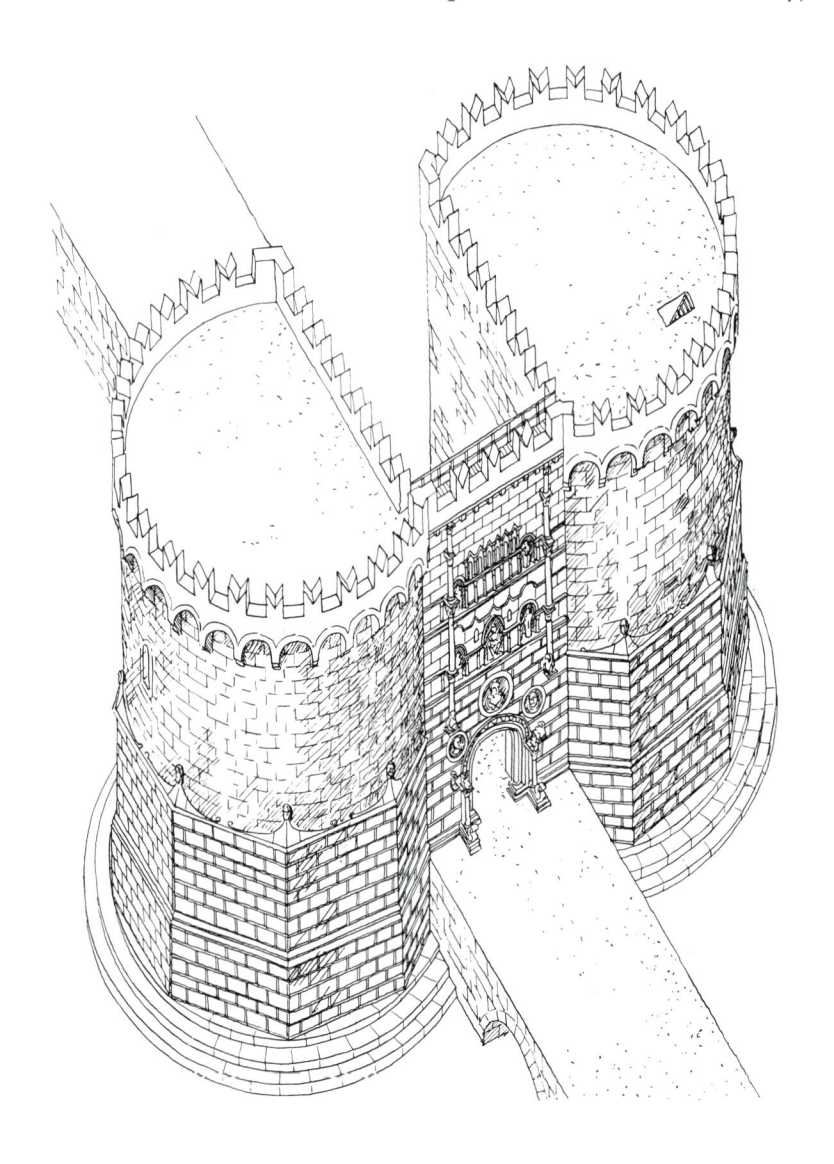

94 Capua, Campania: Attempted reconstruction of the overall layout of the bridge citadel (from Willemsen).

68 *Lines of Development of Hohenstaufen Architecture*

95 East gate section of Qaṣr al-Ḥair al-Ġarbī (second quarter of the eighth century A.D.), Syria. Reconstruction in the museum of Damascus.

96 Torso of the emperor's statue from the bridge citadel. Museo Campano, Capua, Campania.

here as well, drawings and descriptions from the time prior to the destruction have been preserved. They provide us with a very clear idea of the form and shape of the structure and of the arrangement of the sculptural decoration.

The gate stands on the bank of the Volturno, at the head of a bridge leading into the city across the river. Its significance far exceeded this local function, however. As the boundary between the Papal States and the kingdom ran along here, it acquired the importance of a gate of admission to Frederick's empire.

Between two towers rose a facade 8 m wide, subdivided into three main areas (Fig. 94). The middle one accommodated the statue of the emperor. A plaster of paris cast of the head exists, done by Tommaso Solari (d. 1779) in the period preceding the destruction by French revolutionary troops. A drawing, unknown until recently, in the Vatican Library shows the statue torso with the head intact (p. 20, Fig. 3).[71] This made possible the attempt to reconstruct the statue with the help of Solari's terracotta bust (p. 20, Fig. 4),[72] which was apparently made after the cast (see also p. 19 ff.). Below the statue of the seated emperor, in a circular niche, was a colossal female head set over the apex of the portal. On either side of this surviving colossal head and somewhat lower over the portal there were two smaller round niches with male heads, which have also been preserved and are interpreted as representing judges.[73] The borders around the niche openings carried inscriptions, whose text has been handed down. Out of the mouths of the three figures represented it proclaimed the emperor's message:

"By command of the Emperor I am the guardian of the kingdom! I shall cast into disgrace those whom I know to be inconstant. Walk through safely, he who is willing to live without fault, but let the faithless fear banishment and death in the dungeon."[74]

It is amazing how much classical sculptural style has been preserved or brought to life again in the heads as well as in the surviving robed torso of the emperor statue (Fig. 96). All of the works can be found today in the Campano Museum in Capua.

The facade of the bridge gate is flanked by two towers built with extreme care; their polygonal socles (Fig. 97) rise on circular bases. The socles are octagons that are

The Gate to the Bridge of Capua 69

97 Capua, Campania:
Remains of the two bridge citadel towers.

bent open toward the middle, in the direction of the gateway.[75] Through the mirror imagery of the two opposing socles, the asymmetry is subjected to a new symmetrical arrangement: The gate has a symmetrical axis in the middle of the street that passes through it.

Somewhat above the apex level of the gateway, the masonry of neatly cut, embossed squared limestone blocks changes into regular, cylindrical towers built of well-fitted blocks of dark tuff (Fig. 98), which appear just as neat in the tower interiors. We are confronted once more with the endeavor to embellish a structure with plastic devices instead of superficial ornament, first, by means of the transition from one stereometric form to the other, and second, through the different nature of the materials used. What importance was attached to this change from one stereometric form to the other can be deduced from the painstaking elaboration of the transitional zone, i.e., the area between the corners of the polygon and the cylindrical upper part that rises out of it. The transitions are created by spandrel-shaped elements that make a three-dimensional connection as forms that penetrate from a polygonal pyramid mounted on the prism with the cylinder above it. Here again, the sense of the stereometric nature of the building components becomes quite clear. It determines the exclusively plastic character of this architecture, which knows hardly any superficial ornament — unless it arises from the three-dimensional substance of the edifice itself, for instance in the lines of demarcation between the stone blocks. This principle is not contradicted by the sculptural trappings on the central facade of the structure, which serve no architectonic-decorative function, but rather one that is political-programmatic.

Typological precursors to the form of the bridge gate are again found in the Syrian area: e.g., the portal of the Omayyad castle of Qaṣr al-Ḥair al-Ġarbī, dating from the second quarter of the eighth century, which we have already encountered in connection with the prototypes for the Hohenstaufen four-wing constructions in Catania and Syracuse. A reconstruction of this portal can be found in the museum of Damascus (Fig. 95).[76]

The obvious relationship between this portal from Qaṣr al-Ḥair al-Ġarbī and the gate to the bridge in Capua cannot be accidental. I consider it superfluous, however, to postulate that the emperor saw the former with his own eyes. After all, Frederick II was surrounded by nume-

98 Capua, Campania: Base of the western tower of the bridge citadel.

The Gate to the Bridge of Capua

rous scholars who probably had — hardly different from Norman times — the most varied origins, including Asia Minor.[77] Contacts with architects of the Islamic world, particularly Syria, could well have existed through them.

100
Caserta Vecchia, Campania: Cylindrical residential tower.

7 *Caserta Vecchia*

The interesting zone of transition from polygon to cylinder observed on the Capuan towers is repeated in a similar form on a fortified residential tower located near present-day Caserta Vecchia (Fig. 100).[78] The tower in question is a huge cylindrical structure from the Hohenstaufen period, dated between 1240 and 1250. Presum-

99 Lagopesole, Basilicata: Keep within the citadel.

72   *Lines of Development of Hohenstaufen Architecture*

Caserta Vecchia

ably, it acquired the form it has today under Riccardo San Severino, who became Count of Caserta in 1231. The count had been brought up at the court of Frederick II; he married Frederick's daughter Violante in 1248, thus attaining power and affluence.[79]

The powerful cylindrical tower, bursting with plasticity, forms a counterpart to the square-prismatic towers of Bari or Lagopesole (Fig. 99). It is assumed that this tower was built by the same architects who erected the towers of the bridge gate at Capua. They are identical in the construction of a round tower that emerges from a polygonal base. The different nature of the materials used for the base and the tower is also peculiar to both monuments. The base is made of white travertine, while the cylindrical tower element is done in gray tuff. The same decorative use of varicolored material is seen at the portal, which is adorned with a border of white marble.[80]

A difference exists between Capua and Caserta, in that the substructure in Caserta Vecchia has not eight, but sixteen sides. The geometric basis in both cases is the square. The transition from the 16-sided polygon to the cylindrical form is made by spandrels that correspond exactly to those on the bridge towers at Capua. While they are less artistically shaped in Caserta Vecchia than in Capua, these 'spandrels' at Caserta are better recognizable as being the lateral faces of a 16-sided pyramid, as the photograph reproduced here (Fig. 101) clearly shows: two each of the spandrel triangles forming the 16-sided prism lie on one and the same plane, which is part of a pyramid face.

To complete the overall picture and to provide an idea of the variety of architectonic concepts, we shall consider several other, in some cases later, structures.

101 Caserta Vecchia, Campania: Cylindrical residential tower on a decahexagonal base with the formation of spandrels at the transition from the base area to the tower cylinder. It can be seen clearly on the right that the two outer spandrel surfaces lie on one plane, that of the pyramid face.

8 Prato

In Prato near Florence, far beyond the confines of the kingdom of Sicily, therefore, stands the 'Castello dell' Imperatore' (Fig. 103).[81] The builder was related to Frederick II. Based on the sources available, the citadel appears to

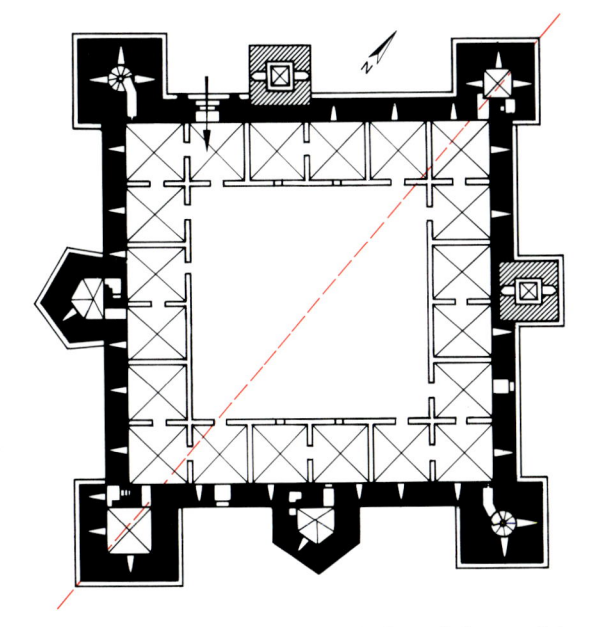

102 Prato, Tuscany: Ground plan of the citadel (the older towers that are integrated into the structure are hatched, line of symmetry is in red).

103 Prato, Tuscany: View of the citadel from the south.

have been completed in 1239. With respect to its ground plan (Fig. 102) it is akin to the 'Byzantine' citadel type exemplified by Augusta (p. 43, Fig. 47). It starts out again from a square; this is repeated in the form of the courtyard that is surrounded on each long side by an even row of six square compartments covered by groined ribbed vaults.[82] Relationships of homeomorphic symmetry exist, going from the basic square, over the square of the courtyard, to the vaulted squares of the gallery. The strong preference for symmetry is discernible in the arrangement of the two older towers that are incorporated into the building on the northeastern and northwestern long sides. They are placed so as to be subject to a diagonal axis of symmetry running from north to south. It is not an exact symmetry, but the efforts made in this direction are evident. This holds also for the polygonal towers on the southeastern and southwestern long sides. Of the four square corner towers, the northern, the eastern, and the western are relatively the same size; the southern one is somewhat larger — these proportions as well conform to the system of the diagonal axis of symmetry. The placement of the corner towers is connected to the layout of Augusta (p. 43, Fig. 47); they project to two sides — in this case toward the southeast and the northwest — more than to the other two, i.e., the northeast and the southwest. Instead of the two hexagonal side towers of Augusta there are two pentagonal

Prato

ones in Prato. Of particular interest is the portal, placed outside of the axis of symmetry (Fig. 104), which bears great similarity to the magnificent portal of Castel del Monte (Fig. 105), although lacking its force and splendor.

As certain as it seems that the group of imperial structures we have discussed — planned and begun within the decade between 1230 and 1240 — can be traced back to the personal ideas of the emperor himself, we should not assume that Frederick II exercised any direct influence on the plan of the citadel at Prato, even if it is just as certain that he was there.

104 Prato, Tuscany:
Portal on the northwest side of the citadel. Its connection to the portal of Castel del Monte is striking, although it does not approach the magnificence and splendor of the latter.

105 Castel del Monte, Apulia:
East portal in the form of a triumphal arch.

Lines of Development of Hohenstaufen Architecture

9 Termoli

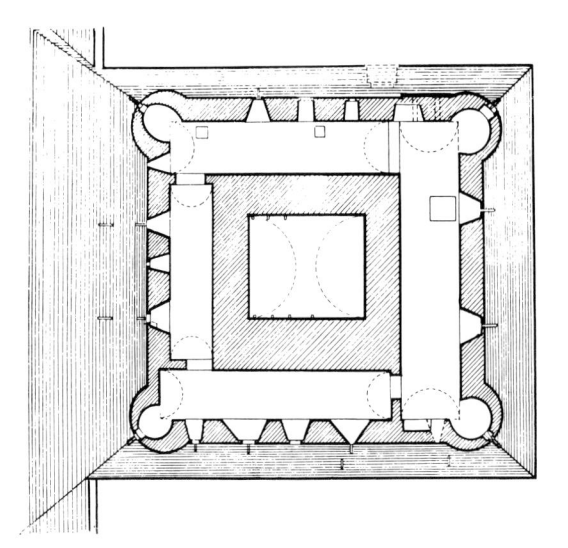

106 Termoli, Molise: Ground plan of the citadel.

Let us turn again to southern Italy and consider a citadel situated in Termoli (Fig. 107), on the northern border of the former kingdom of Sicily, which today belongs to the region of Molise.[83] This extraordinarily interesting structure, which Bertaux counted among the Hohenstaufen buildings on the basis of an inscription referring to Hohenstaufen origin, is an example of the 'donjon type', of Norman provenance. It stands on a truncated rectangular, almost square pyramid, the corners of which are enhanced by cylindrical bastions (Fig. 106); with the sloping walls of the base pyramid they form most striking intersecting figures that might have been taken from a modern textbook of descriptive geometry (Fig. 109). Once more, we are confronted with a building of strict symmetry with four symmetrical axes. Again we distinguish

107 Termoli, Molise: General view of the Hohenstaufen citadel.

77

the square and the circle as elementary figures. The repeatedly noted pleasure in intersecting figures of stereometric elements is carried here to perfection; it remains clearly visible despite later repair of the masonry. The preference for clear geometric structures with a central orientation, the manifold symmetrical relationships, complemented by the play of intersection — all these are already familiar to us as design elements of the monuments

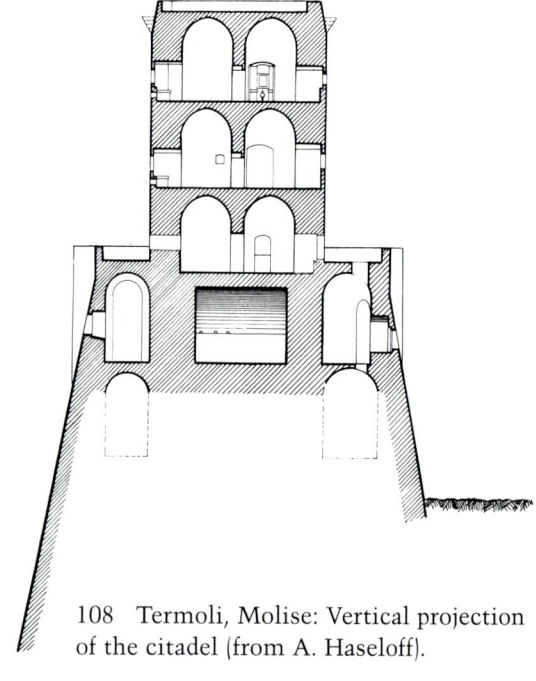

108 Termoli, Molise: Vertical projection of the citadel (from A. Haseloff).

109 Termoli, Molise: Hohenstaufen citadel. The round corner bastions grow out of the pyramid-shaped foundation. Notable are the clear stereometric intersecting figures.

78   *Lines of Development of Hohenstaufen Architecture*

of Hohenstaufen architecture that we have examined. It becomes clear once more that not the two-dimensional superficial ornament but rather the expressive power of three-dimensional plasticity alone is sought in this architecture — an exceptionally important realization. The foundation of this building, with the pyramid level below and the square donjon-type prism that rises above it, has rightly been compared to Lucera.[84]

The form of the citadel of Termoli was often imitated in the region along the Adriatic coast. One of the most austere and well-planned fortresses of the Termoli type lies high in the Abruzzi (1460 m) on a commanding mountain ridge: Rocca di Calascio (Figs. 110 and 111). Damaged by war and earthquakes, the construction consists — as in Termoli — of a foundation in the form of a truncated pyramid with a square ground plan. On this platform arises the donjon as a square prism; the lower story is built of large well-fitted ashlar. At the four corners of the platform, where in Termoli circular bastions grow out of the pyramid edges, stand cylindrical towers in Calascio. They rest on conical bases that begin far below on bedrock (Fig. 111). The combination and intersection of prism, cylinder, and pyramid observed in Termoli is supplemented here by the cone, producing an elaborate interplay of four basic stereometric elements from our repertoire of forms found in Hohenstaufen architecture (p. 108, Fig. 155).[85]

A drawing of Adranò that has not been considered up to this point, from the inexhaustible stores of the French

110 Rocca di Calascio, the Abruzzi: Ground plan of the fortress, which was built southeast of L'Aquila at an elevation of 1460 m (ca. thirteenth/fourteenth century).

111 Rocca di Calascio, the Abruzzi: Partial view of the fortress from the west.

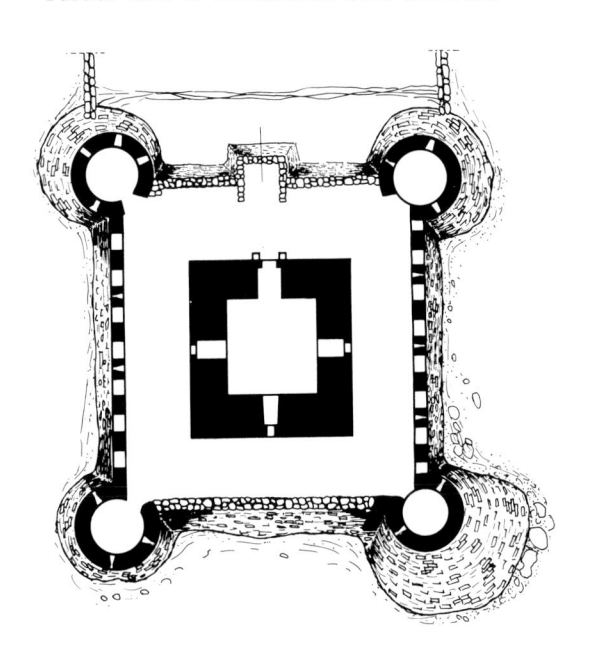

112 Louis Jean Desprez, pen-and-ink drawing of the donjon of Adranò with four towers on the round corner bastions. These towers were added presumably at the end of the fifteenth century and still existed in 1778, the year L. J. Desprez arrived in Sicily (from Wollin 1935, p. 238, Fig. 113).

architect Louis Jean Desprez, takes us from Rocca di Calascio into Norman Sicily.[86] The donjon of Adranò (p. 62, Fig. 81) appears here with four towers on the round corner bastions (Fig. 112); according to local records, these were added during the rule of Count Giovanni Tomaso Moncada (1461—1501) when the donjon was being renovated.[87] They were probably destroyed by an earthquake sometime after 1778 (the year of Desprez' arrival in Sicily). Apparently, however, such a fortress type having corner bastions with fortified towers did exist, as manifested by Rocca di Calascio and shown in a later version at Adranò.

It is tempting to ask whether round towers may once have risen above the round corner bastions in Termoli as well. In any case, a typological relationship is detectable between the donjon of Adranò, sketched by Desprez, Rocca di Calascio, and Termoli.

10 Enna

The magnificent octagonal residential tower in Enna, the old Castro Giovanni in the heart of Sicily (Fig. 115),[88] which unquestionably belongs in the realm of Frederick's Hohenstaufen structures, takes up once again the main theme of this architecture: the octagon — evolved from

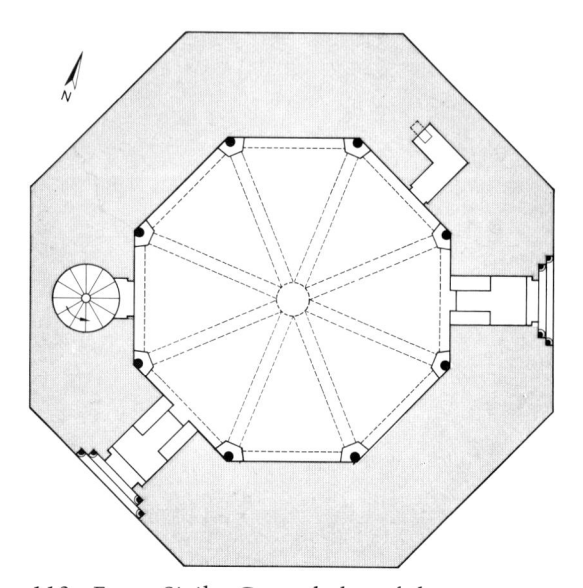

113 Enna, Sicily: Ground plan of the eight-sided 'Torre di Federico'.

80   *Lines of Development of Hohenstaufen Architecture*

two squares rotated toward each other by 45°. It is a donjon-type building. The octagon with a diameter of 17 m dominates here with an ingenuity surpassed only in Castel del Monte. The ground plan of the three interior rooms, one above the other, is also octagonal (Fig. 113). The tower, built of golden yellow limestone in a marvelous squared-stone technique, is surrounded at a distance of about 23 m by an *octagonal* wall (Fig. 114). The homeo-

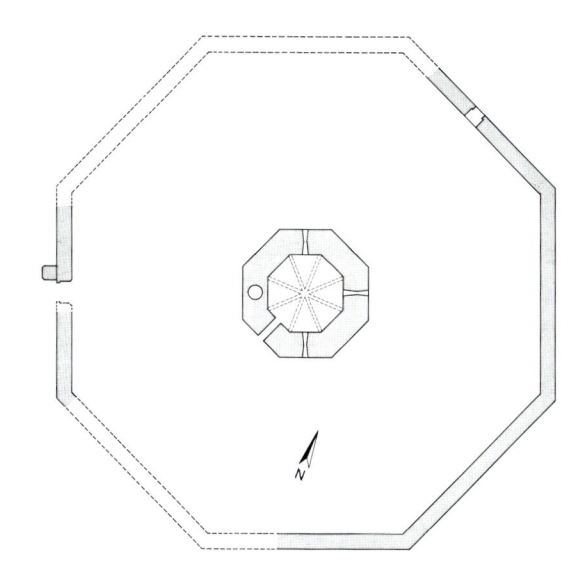

114 Enna, Sicily: Plan of the 'Torre di Federico' and the octagonal wall of the surrounding courtyard.

115 Enna Sicily: General view of the 'Torre di Federico'.

Enna 81

116 Enna, Sicily: Window on the first floor of the 'Torre di Federico'.

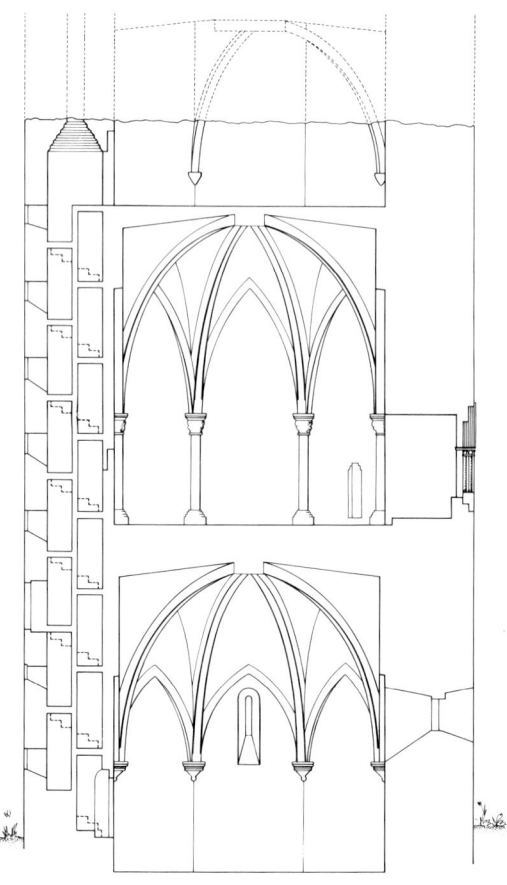

morphic symmetry of three octagons, one inside the other, reminds one of Castel del Monte. The workmanship of the wonderful, evenly cut squared-stone masonry brings to mind another Sicilian structure: Castel Maniace in Syracuse — the techniques used for the relieving arches over the windows are also similar (compare Fig. 116 with p. 38, Fig. 33).

The fondness for strict stereometric form impresses here in southern Italy with exactly cut edges that set off the surfaces, making them even more handsome and pure (Fig. 115). No decorative ornament interrupts the smooth masonry of the sharp-edged prism, other than two windows (Fig. 116), whose three-dimensional graded depth is another reminder of the still more richly formed window in Castel Maniace (p. 38, Fig. 35).

Attention has justly been called to the relationship with the ruins of the grand octagonal donjon of Egisheim in the Alsace. Each side of the octagonal wall there measures 12 m. The construction is dated generally to the second half of the twelfth century; in view of its condition, this seems venturesome. In this connection, W. Krönig cites the fortress of Steinsberg (Fig. 118) in Baden,

117 Enna, Sicily: Longitudinal section through the 'Torre di Federico' with its three stories.

82   *Lines of Development of Hohenstaufen Architecture*

118 The ruins of Steinsberg in Weiler, district of Sinsheim, Baden: The impressive construction from the late Hohenstaufen period was first mentioned in 1109.
The octagonal fortress is built of handsome bossed squared stones. Over the square base stands a multistory residential structure with a circular ground plan.

with an octagonal keep "dating from the late twelfth century",[89] while E. Adam places it in the first half of the thirteenth century.[90] Be that as it may — further careful research is necessary to clarify the temporal relationships. Subjects of investigation should be whether the octagonal keep[91] made its debut earlier in the north than in the south, and whether it played a role in the development of the Hohenstaufen style similar to that of the Norman donjon. In addition to Egisheim, the ruins of the Burgstall in Guebwiller, Alsace, appear to go back to an octagonal structure. Christian Wilsdorf points out that the presumed builder of the stable, the abbot Hugo, in addition to his function as the abbot of Murbach (1216–1236), was also an advisor to Frederick II, whom he often accompanied to Italy.[92] He was one of those who organized the crusade of 1228/29, in which he himself participated — the abbots of the noble collegiate church were princes of the empire! Finally, the castle of Neu-Leiningen in the Rhineland-Palatinate must be mentioned, erected 1238–1241 by Count Frederick III of Leiningen, a follower of the House of Hohenstaufen.

It must not be overlooked, however, that such regular, strictly octagonal structures in the north should be regarded as unusual and are more or less restricted to the vicinity of the Hohenstaufen dominions, where Frederick repeatedly spent time.[93] All this would indicate that the style of architecture bearing the stamp of Frederick himself spread from its origin in his kingdom of Sicily.

11 Osterlant

by R. Spehr

The ruins of the "Wüstes Schloß Osterlant" (deserted Castle Osterlant; Fig. 119) still stand 10 m high. It was an 'oriental' palace in the eastern part of the kingdom, just west of the small Saxon town of Oschatz, in a wood that came under the domain of Margrave Dietrich von Meißen as an imperial forest in 1210. In 1991/92 I directed an archaeological excavation that revealed the entire ground plan of the structure (Fig. 120) and determined that it had been intended as the hunting lodge and summer residence of the Margrave von Meißen (1198–1221).[94] The symmetrical four-wing layout in a square with sides of 46 m surrounds a square interior courtyard (20 × 20 m) that was not accessible from the outside and was probably conceived as a garden and a quiet place. The overall construction repeats the cloister square of the Cistercian monasteries, as well as the successors to the Roman and Byzantine four-wing citadels of the Mediterranean area; it even reminds one strongly of the caravansaries and Muslim desert palaces of the Orient. Our palace was possibly laid out according to the stars. Its main facade (with an axially ordered, elevated entry portal) in the form of a "Roman portico villa with corner projections" is situated near an artificial pond; it was even connected with this by means of a short, navigable canal. The outer walls and corners of the three-story castle were symmetrically arranged, with 20 powerful, bastion-like projections rising to the roof and a surrounding ornamental ledge. All rooms in the basement and on the ground floor were possibly intended to be ceilinged with groined vaulting; in any case, the full number of corresponding wall and corner pillars were excavated. The upper two stories, reached via two stone spiral staircases set into the walls of two projections, had flat ceilings. The lower story, accessible only from within the upper story, had narrow slits for windows. The upper stories could be entered through large, arched doors in blind mirror niches and were illuminated by wide-open, arched windows decorated with columns. Given the absence of any roof tiles among the ruins, the question arises as to whether the Mediterranean model determined even the type of roof (clay-floor flat roof), or whether this also, like much else, remained unfinished.

The palace was built over a small valley like a bell, in order to incorporate a spring. This was enclosed in the eastern corner of the inner courtyard in a square spring house, and in a manner quite splendid for a secular building: The round interior with a diameter of 6 m, made octagonal by contoured wall columns, was reached via two angular steps set into the wall. The round basin (3 m in diameter) was carefully fashioned out of six steps of bright green porphyry and had a contoured octagonal rim that bordered the narrow perimeter and was ornamented

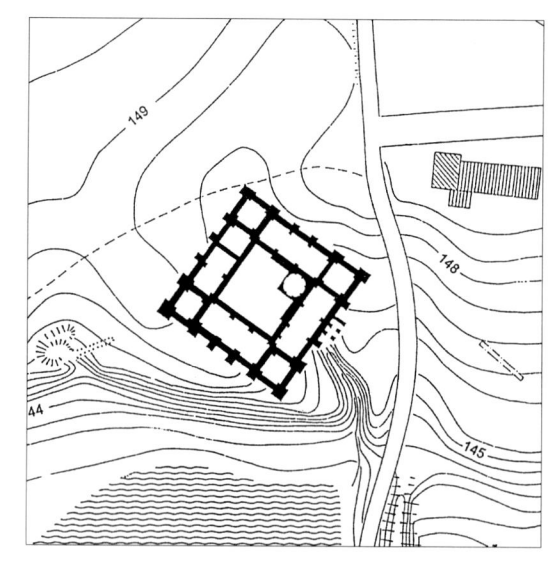

119 Osterlant castle near Oschatz (Saxony), begun in 1211/12.

120 (p. 85)
The ground plan of Osterlant castle, uncovered by R. Spehr, which was erected over the course of a narrow stream. The plan corresponds to the Hohenstaufen four-wing structures in the Mediterranean area previously presented here. The outer square has sides of 46 m, the inner courtyard is a square with sides of 20 m. In the eastern corner of the courtyard is a square spring house with a richly appointed octagonal basin. A round pool, 3 m in diameter, is carefully decorated with green porphyry and accessible from outside via two steps set into the wall. The house is situated above a strong spring. Overflowing water is conducted into the streambed outside via a concealed spillway.

84 *Lines of Development of Hohenstaufen Architecture*

with pedestals (Fig. 121). A cleverly concealed spillway carried excess well water into an underground canal system that had been laid in the foundations of the palace at the beginning of construction. It was possible for the spring room to display its architectonic and colorful decoration

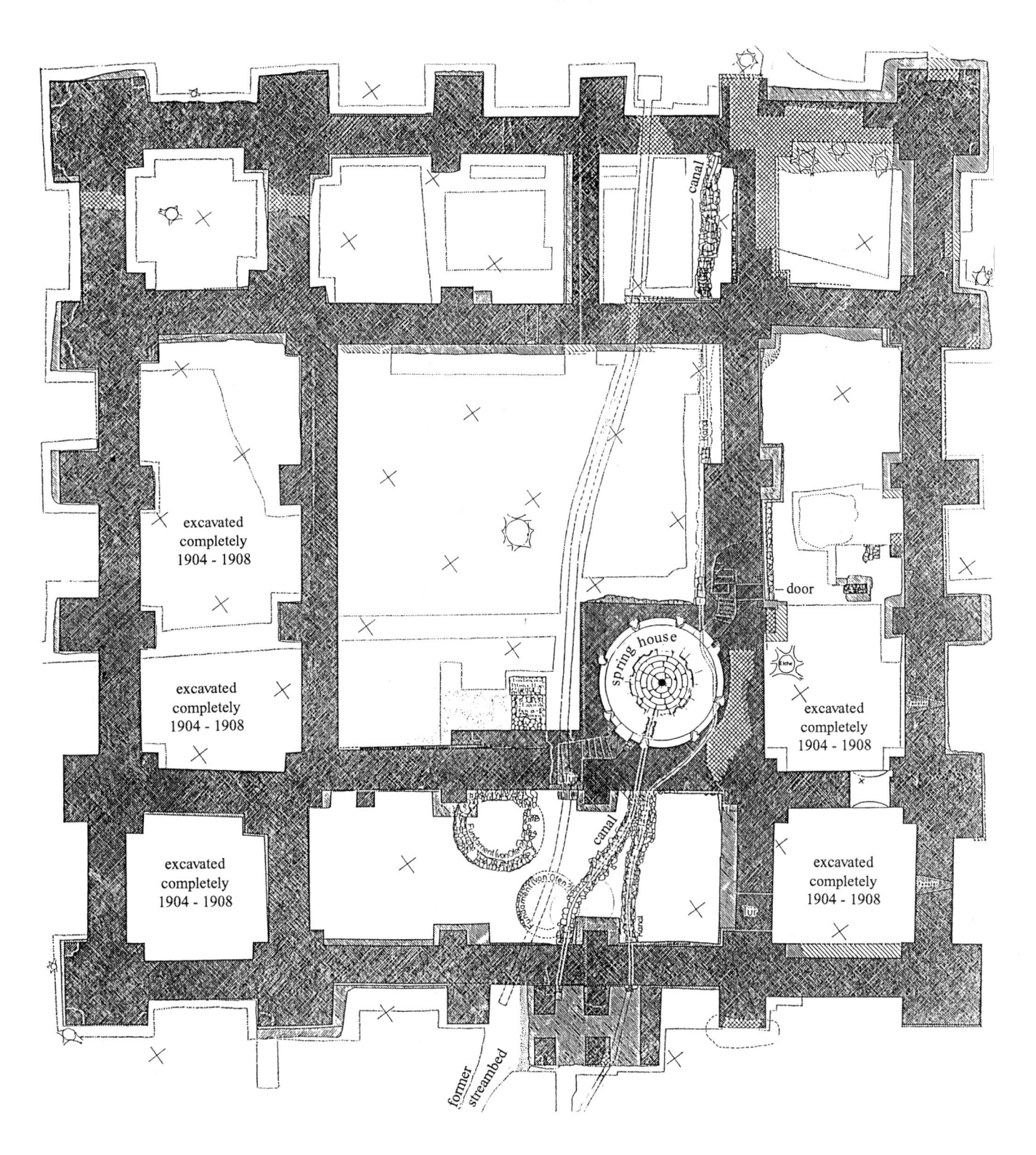

Osterlant 85

121 Osterlant, Saxony: Basin of the spring house. The careful workmanship is recognizable. The use of colored stones gave the entire structure a splendid appearance.

(white sandstone, red and green porphyry) only in daylight, which is why, as there are no windows in the lower story, we would like to reconstruct an upper story interrupted by an arcade. Even in the ruinous condition in which it was found, the room possessed the dignity of a baptistry. Its purpose can really be found only within the ranks of a courtly, fraternally organized society of knights and diners with the margrave as 'primus inter pares'.

According to dendrochronological dating, lumber that was built into the foundations was felled in the winter of 1211/12. This commencement of construction corresponds well with the rich find of archaeological material and with the architectural-historical dating. The palace was never completed. The entire foundation ground plan and two wings of the building were finished, without the possibly planned vaulting of the lower story; a third wing was made inhabitable in 1221 with provisional building methods; the fourth wing remained only a foundation up to the present day — only a wall was built to enclose the courtyard there. The lack of an exterior wall and a moat around the indefensible castle-like house is perhaps due to the fact that it was left uncompleted following the death of the owner in 1221. The findings show evidence that the house was used up to the second half of the thirteenth century; gradual ruin led to abandonment of the house, probably prior to the turn of the century. Not a single written source reports the construction and the existence of this building; in 1379 it was mentioned as a deserted stone house ("vor dem wüsten Steynhuse").

The origin of this palace, unique to central Europe, lay in the fascination that its builder, Margrave Dietrich, brought back from the Holy Land in 1198. The 'oriental appearance' of the structure may have been guided by the minnesinger Heinrich von Morungen,[95] who was a friend of the margrave until his death and probably undertook, together with Bishop Konrad von Halberstadt, a crusade to Byzantium and a pilgrimage to Palestine between 1201 and 1205, and perhaps even reached the grave of the apostle Thomas in Edessa (Mesopotamia). The margrave possibly engaged the lodge of the cloister Walkenried to carry out the construction, and perhaps building experts from the abbeys of Lauterberg and Morimond were called in as well; Italian master builders were apparently not involved.

In 1210, Margrave Dietrich von Meißen was also enfeoffed by the emperor Otto IV with the Ostmark (eastern march), i.e., Osterland. In the previous year he had been accepted into the Brotherhood of Knights of the Emperor. One year later he began with the building of Osterlant palace as a house for hunting parties and banquets for his own brotherhood of knights. Construction was apparently interrupted by the senior civil servants' rebellion of 1216/17; when the house was not even half finished, death put an end to the ruler's dream (1221).

Margrave Dietrich was unusually closely associated with Frederick II during the latter's stay in Germany: In the course of the first expedition against the emperor Otto IV in 1213 (August–October) he sided with the king and served as a witness in the military camp on October 19, 1213. In June of 1214 he attended a meeting of dignitaries at the court in Eger (Cheb); in the winter of the same year he delivered an important message to Frederick at Metz, then accompanied the king from there to Erfurt, Altenburg, and Halle (up to February 1215). In September 1215 he was again with Frederick II in Würzburg. The most intensive encounter of the two rulers took place from September to November of the year 1216: From Altenburg in the Upper Palatinate the two rode together to Leipzig, where the margrave was guaranteed rule of the city by virtue of royal authority. In the fall of 1217 the king and the margrave fought together once more against the emperor Otto, subsequently traveling from Braunschweig and Quedlinburg to the court meeting in Altenburg (November 8, 1217). The two met for the last time at a court meeting in Erfurt in July 1219. Margrave Dietrich was married to Judith, a granddaughter of Frederick I's half sister Judith (i.e., a great-aunt to Frederick II).

Hohenstaufen architecture, whose characteristics show the hand of the emperor, had a remarkable influence. The Sicilian type of citadel with a square ground plan of multiple symmetry and with round corner and/or side towers had unusually long-lived and wide-reaching effects. Some examples are the citadels of Lahr in Baden (Fig. 122)[96] and Neu-Leiningen in the Rhineland-Palatinate,[97] the fortress of Morges in Switzerland (Fig. 123),[98] Bodiam Castle (end of the fourteenth century) in England,[99] or Montealegre (fourteenth century) in Spain.[100] Further examples could be added to the list,[101] down to the castle of Chambord on the Loire,[102] in which this type even appears twice, a smaller square within a larger rectangle (Fig. 124).

122 Lahr, Baden: Reconstruction of the Tiefburg from the middle of the thirteenth century (from K. List).

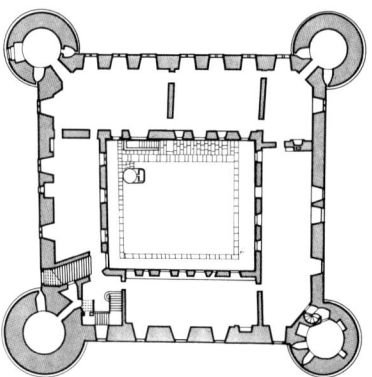

123 Morges, Switzerland: Ground plan of an almost square fortress with round corner towers from the thirteenth century.

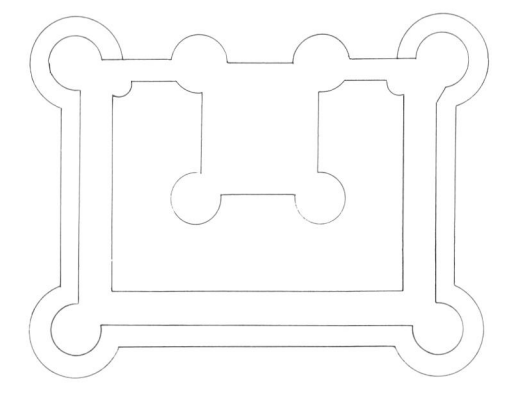

124 Chambord on the Loire: Schematic ground plan of the castle begun in 1519 by Domenico da Cortona and other architects.

12 Castel del Monte

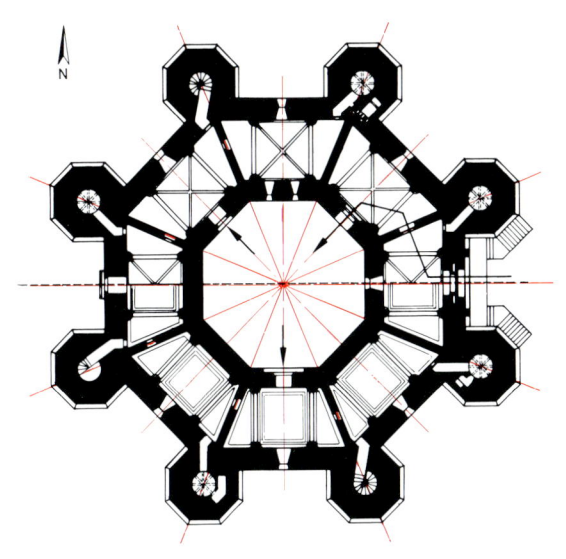

Let us now turn to the most beautiful and most important building by Frederick II, the 'crown of Apulia', which always fascinates people anew, and the origin and interpretation of which are still the subject of controversy. Frederick's architecture reached its zenith in Castel del Monte.

The ground plan (Fig. 125) is based on an octagon constructed from a square rotated 45°, which forms the core of the building. This in turn surrounds an eight-sided court, and an eight-sided tower is set on each corner of the basic octagon. As one approaches, the citadel glows from the distance like a fairy-tale apparition and, as one draws closer, it unveils ever more impressive shapes — particularly when the sun divides the building into areas of light and shadow and reveals its sharp contours (pp. 6/7 and Fig. 128). The aesthetic fascination lies in the effect of the bundles of eight-sided prisms that grow out of the ground, comparable to a crystal. This crystal-like appearance gives the building a natural strength. It is true that we will find here again the features of Hohenstaufen architecture that were characteristic for the end of the emperor's reign and that we discovered in the Sicilian citadels and in Lucera

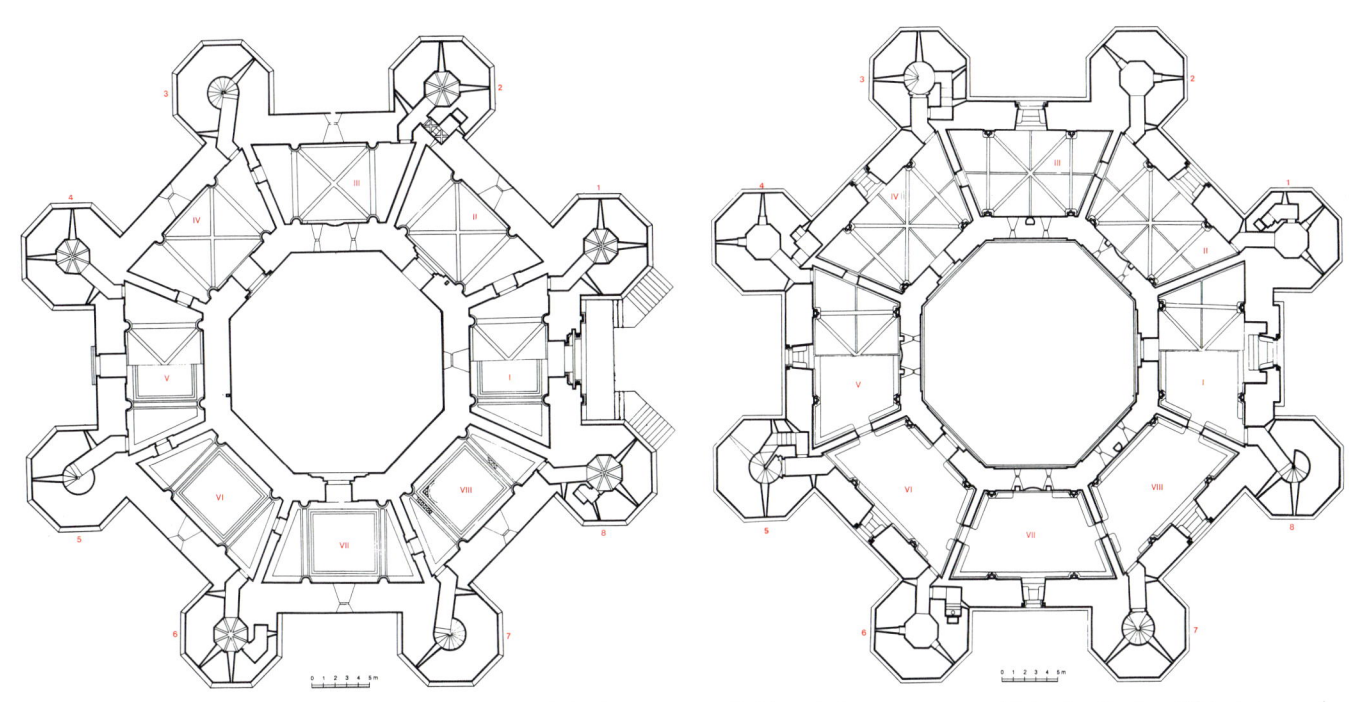

125 Castel del Monte, Apulia: Plan of the ground floor with lines of symmetry.

126 and 127 Castel del Monte, Apulia: Plans of the lower story (left) and the upper story (right; both after Chierici 1934).

and Capua; nevertheless, the building is without example — and this fact is especially surprising: there is no fortress and no castle belonging to the period of the Middle Ages in Europe that could be regarded as a forerunner of this building concept, notwithstanding some particular features in Syracuse and Lucera that might be considered to anticipate individual designs realized at Castel del Monte. Despite the lack of a prototype, the structure is mature and embodies a creative high point. Nothing about it is a groping attempt, nothing is preliminary; rather, every aspect appears in itself to be consistent and consummate.

Where did this collection of forms originate? Where does it have its roots, and what is the source of the intellectual power that shaped it? The origin of related

128 Castel del Monte, Apulia: View from the southeast.

129 Pisa, Tuscany:
Pediment decoration showing Islamic influence on the west facade of the cathedral (eleventh century).

130 Troia, Apulia: Islamic ornament in a window niche on the eastern side of the nave of the cathedral (twelfth/thirteenth century).

individual forms has been successfully researched. There is no question that the Hohenstaufen architecture of southern Italy is connected with Gothic style — in particular, that of Burgundy and the Champagne, conveyed by the architecturally versed Cistercians, with whom Frederick II felt a special bond. At the same time, naturally, this Hohenstaufen architecture is in the Norman-Arabic and Byzantine tradition of the kingdom of the Two Sicilies. Lines can also be traced to the homeland of the Hohenstaufen in Germany. Finally, the heritage of 'Magna Graecia' should not be forgotten — quite surprisingly, geographically identical with the area of the Sicilian kingdom, which was marked for many centuries by the Greek tradition of the 'classical' period.

Do these general statements suffice for us to grasp what is special and unique about this architecture, which, despite all the affinities mentioned, resists the customary stylistic classifications? In our consideration of the buildings characteristic of Frederick's architecture, in particular the four-wing constructions Castel Maniace in Syracuse, Castel Ursino in Catania, and Lucera, we have identified clear links to Byzantine and Arab-Islamic architecture. This holds for a large number of decorative elements which found their way, via lively trading with the Islamic-occupied coasts of the Mediterranean region, to Italy and Sicily; the latter was itself under Islamic rule for more than two centuries (Figs. 129 and 130).[103] Moreover, we have become aware of forms and design principles of Hohenstaufen buildings that have their roots in the Byzantine, but especially the Arab-Islamic world and its mathematical-geometric tradition.

Castel del Monte elucidates further what the great citadels discussed up to this point have taught us to appreciate: This development, proceeding briskly forward

131 Castel del Monte, Apulia:
Detail of the left side of the portal pediment.

and hurrying toward its apogee, is inconceivable without a creative personality in the background. The text by Riccardo da San Germano that has been handed down concerning the building of the bridge citadel of Capua testifies impressively to the emperor's active participation in designing the structures he commissioned (cf. p. 67f). Frederick II, who attached great importance to the effect of imperial might and dignity, clearly recognized the role architecture played in the representation of his sovereign power. As an architect the emperor displayed strength in setting standards, originality of thought, and creative enterprise. He moved within the field of tension of the varied traditions in the Mediterranean region, which grew together under his influence into a new style that was expressed to perfection in Castel del Monte.

Let us now deal with the ground plan in detail. Around a central eight-sided court, in the middle of which, according to old records, an octagonal fountain once stood,[104] rises a two-story, eight-sided building. In the middle of each trapezoid room is a square area ceilinged by the familiar Cistercian ribbed groined vault. Adjoined to each of the eight outer corners of the centrally planned building is an octagonal tower; the two sides that are directed toward the center of the building disappear into the body of the large octagon (Figs. 132 and 133).

Only six sides of each eight-sided tower are visible; the tower walls that meet the large octagon are perpendicular to it. The main portal — resembling a triumphal arch — is placed in one of the eight wings, facing to the east (Figs.

132 Castel del Monte, Apulia:
The incorporation of the corner towers into the main octagon can be seen clearly here.

92   *Lines of Development of Hohenstaufen Architecture*

131 and 134). The dark, glowing color of its *breccia rossa* (coralline breccia) contrasts stongly with the rest of the citadel. On the right and the left, perrons lead up to the portal. The building is flat on top, and today it is no longer possible to determine how high the wall cornices were and how much higher than the main building the corner towers may have been.[105]

The trapezoid rooms, all of the same shape, were richly appointed; the walls of the lower story were faced with *breccia rossa*, those of the upper story with marble. The border of the facing, which extends to the base of the vault, is formed by a continuous narrow molding that elegantly runs around the door and window frames. Along the walls of the upper-story rooms are stone benches of

133 Castel del Monte, Apulia: Incorporation of the corner towers into the main octagon. The ridge that runs around the building indicates the division between the two stories. The tower bases are also clearly visible here.

Castel del Monte 93

94   *Lines of Development of Hohenstaufen Architecture*

134 Castel del Monte, Apulia: Portal in the form of a triumphal arch.

light-colored travertine. In two rooms of the lower story and in five of the upper story there were tall, slender fireplaces. They are badly damaged; in some cases only vestiges remain (Fig. 135). Doors and windows are richly decorated with *breccia rossa*, particularly the portals of varying design facing the courtyard (Fig. 143). The most splendid is the portal on side VII of the court (p. 106, Fig. 153). The rooms of the lower story, which has much thicker walls on both the court side and the outside than those of the upper story, are much more simply decorated, and, owing to the meager amount of light that enters, they appear gloomy compared with the cheerful rooms of the upper story, where light pours through doors and windows.[106]

All of the rooms on both stories are dominated by a square Cistercian-type ribbed groined vault; on the lower floor these rest on heavy, round, single columns made of

135 Castel del Monte, Apulia: Hall II of the upper story with remains of the fireplace. The stone benches running around the walls are clearly visible.

Castel del Monte   95

136 Castel del Monte, Apulia: Hall VI of the lower story with the beginning of the crossed ribbed vault. Left, the base of a rib for a (nonexistent) adjacent yoke. The thickness of the planned incrustation can be deduced from the ridges of *breccia rossa* next to the columns.

breccia rossa (Figs. 136, 137, and 139), while on the upper floor slender, three-column bundles made of beautifully patterned light-gray marble convey a much lighter impression, which is enhanced by the finely chiseled capitals (Fig. 135 and p. 110, Fig. 156). The bundled columns and capitals are a clear reminder of Castel Maniace (see p. 40 ff.). The visible ribbed groined vaults had no static but only a decorative function; they were placed in front of the supporting vaults without actually being attached to them (Fig. 138). The square shape of the vaults is copied on the floor by corresponding squares bordered by strips of white marble. On six sides of the lower story there are round-arched windows framed with *breccia rossa*. For reasons of safety, the window openings lie quite high above floor level and are slanted strongly to all sides, but especially toward the floor.

137 Castel del Monte, Apulia: Hall VI of the lower story. Column base of *breccia rossa*. Five sides of the octagonally designed column base are visible.

96   Lines of Development of Hohenstaufen Architecture

138 Castel del Monte, Apulia: Crossed ribbed vault in hall IV of the lower story.

139 Castel del Monte, Apulia: Hall VI of the lower story with the door frame of the walled-up side entrance.

140 Castel del Monte, Apulia: The only three-part window between the second and third towers with a small bifora. It is called the 'Andria window', because it faces the town of Andria, loyal to the emperor, where two of the emperor's wives were interred.

The upper story displays very artistically designed two-light mullioned windows in rectangular wall frames which do not lie exactly on the axes of the lower-story windows. The frames consist of graduated jambs and set-in columns, plastically fashioned of white marble and *breccia rossa* — similar to the portal and windows of Castel Maniace. Only between the second and third towers is there a much wider mullioned window with three lights (Fig. 140) with a small biforate window set above it that takes the place of the rosette above the other windows. This window is said to have been especially richly appointed, as it looks down upon the city of Andria, which the emperor called 'Andria Fidelis' because it was loyal to him. In addition, the dead wives of Frederick II, Isabella of England and Isabella of Brienne, had been laid to rest in the crypt of the city's cathedral.

Missing from the windows on the upper story are the middle columns made of marble; in the second half of the eighteenth century they were taken to the park of Caserta, designed by Luigi Vanvitelli.[107] Between the fifth and the sixth tower there was a side entrance. It is now walled up, but the door frame can still be seen from within (Fig. 139).

The inner court, which contained cisterns beneath the pavement, is decorated with slender arcatures. Older sources indicate that an eight-sided fountain stood in the middle of the court.[108] The pavement that looks so cheer-

141 Castel del Monte, Apulia: View from room II through the portal leading into the inner court. Note the three-dimensional gradation of the door frame.

98 *Lines of Development of Hohenstaufen Architecture*

less today was presumably relieved by plants, as had been usual in such *impluvium*-like courts since Roman times. Above the northwest portal of the inner court (fourth facade) there is a sculptured torso in poor condition that is believed to be the remains of a male figure (Figs. 142 and 143). It has been interpreted as a horseman dashing out of the wall. On the second facade of the court to the left of the arcature, high up and difficult to see, is a probably classical relief that is also badly destroyed (Fig. 144). Its placement is curious, as it is nearly hidden from admiring glances.

The interiors of the eight corner towers are artistically and appropriately decorated. Towers one, two, four, and eight, accessible from the lower story, repeat the exterior octagonal form in the interior rooms, which are topped by octagonal ribbed vaults (cf. Fig. 147). In the third, fifth,

142 Castel del Monte, Apulia: Torso of a horseman over the portal of the fourth facade of the inner court.

143 Castel del Monte, Apulia: Northwest portal of the inner court (fourth facade).

Castel del Monte

144 Castel del Monte, Apulia: Badly destroyed, possibly classical relief on the second court facade on the upper story next to the arcature.

and seventh towers cylindrical spiral staircases lead to the upper story. Via the winding staircase in the fifth tower one can climb from the fifth hall up to the roof. Whereas the adjoining rooms in the first and fourth towers served no recognizable specific purpose, sanitary installations are found in the rest. In the lower level of the second tower are the most lavishly appointed sanitary facilities.

The vault over the spiral staircase in the seventh tower is six-celled (Fig. 148). The cylindrical shape of the room in which the newel rises allowed the construction of a hexagonal vault by division of the enclosing circle with its radius. A special feature of this six-part cupola is six crouching atlantes with very expressive, in some cases grotesquely shaped masks. Similarly grotesque heads rest on the consoles of the three-ribbed vault in the third tower (Fig. 149), the 'Falconer's Tower' (*torre del falconiere*). Following provisions in the emperor's book *On the Art of Hunting with Birds*, an artificial eyrie was constructed here for the breeding of young falcons. Only two of the vault's severies are closed; the third is open, and with the help of a ladder one gains access through the opening to a dark room, the 'artificial eyrie'. From here there was a separate entrance to the continuation of the spiral staircase, leading further upward to the roof. All this is painstakingly arranged, as are the sanitary rooms with their washing facilities that were probably fed by cisterns. Light entered the towers only through narrow

145 Castel del Monte, Apulia: Hall IV of the upper story, opus reticulatum on the internal wall to the next room above the wall bracket.

100 *Lines of Development of Hohenstaufen Architecture*

slits that were slanted inward like embrasures. They are very cleverly placed to provide for light and the circulation of air. Viewed from the outside, they do not interrupt the uniform picture of the towering walls in any way.

The sculptures mentioned and the work on the capitals and bosses of the ribbed vaults are of astonishing quality. The best master craftsmen must have collaborated on the sculptural decoration of Castel del Monte.

The individual rooms are accessible from the entrance only via the central court in nearly unavoidable sequency — an architectonic measure that probably met the need for security.

As regards the intended purpose of each given room we are left to speculate. Thus only one room on the upper floor, situated on the portal side above the entrance, has been designated from time immemorial as the throne

146 Castel del Monte, Apulia:
Arcature and door-like window in the fourth facade of the inner court on the upper story.

147 Castel del Monte, Apulia: Eight-part vault with very pronounced ribs in the lower story of the fourth tower. The tower interior forms an eight-sided prism.

148 Castel del Monte, Apulia: Six-part vault over the spiral staircase of the seventh tower with crouching atlantes. The tower interior forms a hollow cylinder.

Lines of Development of Hohenstaufen Architecture

149 Castel del Monte, Apulia:
Three-ribbed vault in the third tower,
the 'Falconer's Tower' (*torre del falconiere*).

room. It is interesting that it lies at the end of the long walk through the rooms that precede it to the south and has no door to the neighboring room to the north. It is distinguished by higher steps to the window seats, owing to the considerably greater thickness of this portal wall, accommodating the mechanical equipment for the portcullis, which demands a great amount of space. It does not seem totally unfounded to ascribe to it the function of a throne room, given that it is situated over the front entrance with its windows facing east. At the vertex of the groined vault there is a boss decorated with an animated mask.

In hall VIII of the lower story a remnant of the mosaic floor is still preserved; it is shown here in Fig. 150 and its ornamental structure is completed in a drawing (Fig. 151). Here, too, a system of forms becomes visible that reveals Islamic influence. The mosaic creates an interesting pattern with a hexagon-based structure. The hexagons of variously shaded colors are not arranged like a honeycomb; that is, they do not abut with their sides but rather with their points. This creates triangular spaces between

Castel del Monte 103

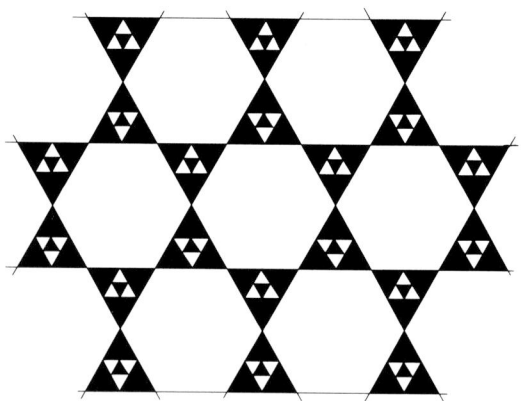

them, and their division into further smaller triangles produces a pattern known in fractal geometry as a 'Sierpinski pattern'. It is easy to imagine that this pattern can be continued into infinity, i.e., that the pattern becomes more and more dense, which is why we also speak of 'Sierpinski density'. In any case, here as well, the pleasure taken in geometrically unusual decorative solutions is evident.[109] As a point of interest, there are other places in Italy where a fractal is clearly represented. One is found in a mosaic on the floor of the cathedral of Anagni (1104). The authors of a publication entitled *Fractal Forms*[110] designate it as a 'Sierpinski gasket fractal' in the fourth order of iteration and believe that it may be the oldest man-made fractal object (Fig. 152).

Outwardly, the structure is distinctly divided into a base area and an upper area; the separation of the two stories is indicated by a torus running around the entire building (Fig. 133). This appears to anticipate Renaissance architecture, in which the dividing line between palace floors is also outlined on the exterior. At the same time, it testifies to a profound understanding of plasticity on the part of the building's creators; while allowing of no linear superficial ornamentation, they aspired to make the building visible as a unit composed of individual, organically arranged parts. The overall design is dominated by one basic chord: the regular eight-sided prism, which is first sounded in the massive outer walls of the central octagonal building, returns in the form of the

150 Castel del Monte, Apulia: Room VIII on the lower story. The floor mosaic displays a hexagonal basic structure.

151 Castel del Monte, Apulia: Schematic drawing of the floor mosaic. It forms a so-called fractal (see p. 188, Fig. 255).

104 *Lines of Development of Hohenstaufen Architecture*

152 Anagni, floor mosaic in the cathedral (1104) with a fractal in the form of a Sierpinski gasket in the fourth order of iteration. This is considered to be possibly the oldest man-made fractal object (from Gruyon and Stanley 1991). Like the one in Castel del Monte (Fig. 150), this one is based on the triangle or hexagon. It is interesting that in our analysis of the geometrically determined ground plan we also encounter the construction of 'fractals' (see p. 186, Fig. 253).

inner court, and is taken up as an eightfold arpeggio by the corner towers. Twice four instances of axial symmetry and eight rotation symmetries can be found. In addition, there are the homeomorphic symmetries — so impressive for the aesthetic sense — of the large citadel octagon with the octagon of the court and the octagons of the eight corner towers. The overall plan of the structure is governed by these symmetrical relationships.

In our consideration of other Hohenstaufen structures we have already seen that there are practically no decorative elements of a linear nature. This does not lie in the fortress-like character of the buildings but is a principal part of the architectural concept. Another conspicuous feature is the three-dimensional quality that is accentuated by the stereometric relationships. Broad outer walls with sharp-edged borders, neat squared-stone masonry with smoothly cut stones — occasionally ashlar bosses: these are the techniques with which this 'geometric architecture' was realized.

Bertaux believed the architect of Castel del Monte to have been Philippe Chinard, and investigated and clarified his genealogy.[111] This is not the place to examine this claim in detail, the less so as Bertaux's principal concern was to prove that the Hohenstaufen architecture in southern Italy was purely 'French', with its roots in Burgundy and the Champagne. Without detracting in the least from Bertaux's generally excellent work, let it be said that the view he takes in this respect is not tenable.[112]

It culminated, among other things, in the assertion that not only the Cistercian artisanship, but also the ground plan of Castel del Monte itself, and thus the overall concept of this building proved to be of purely French origin. He justified this by pointing out that it could easily be traced to the ground plans of the polygonal chancels of St. Remi in Reims (1170/80) or Nôtre-Dame in

153 Castel del Monte, Apulia: Portal on the south side of the inner court (seventh facade).

154 Burgos, Spain: Portal of the cloister Las Huelgas.

Châlons-sur-Marne by duplication of their mirror image. We shall see that the ground plan of Castel del Monte has a different source.

What Bertaux convincingly worked out is the origin of the Cistercian stonemasons' lodges of lower Italy in the centers of the early Gothic period, with their technique of ribbed groined vaulting, their type of capitals, and certain forms of 'interior design' that emerged therefrom. There is no doubt that the interior decoration of Castel del Monte and some particular details, such as the design of the portal, show the influence of the Cistercian lodges. However, this says nothing about the author of the overall concept of this Hohenstaufen citadel. The ingenious ground-plan design, the fundamental shape based on the square with its manifold symmetries, the complex mathematical relationships, and finally the stress on three-dimensionality have other roots. They bear the stamp of the architecturally versed imperial builder.

It is not only the overall ground plan that has an exceptional origin, however; the inner court with its portals is also not purely French-Gothic in style and spirit.[113] There are signs of influence from other sources here that should be analyzed in detail. One example is the similarity between the two-legged, roof-shaped pediment over the arch of the portal on the seventh courtyard facade (Fig. 153) and the portal of Las Huelgas cloister near Burgos (1175; Fig. 154), which is assigned to the Spanish-Moorish style of the Mudéjares.

Castel del Monte 107

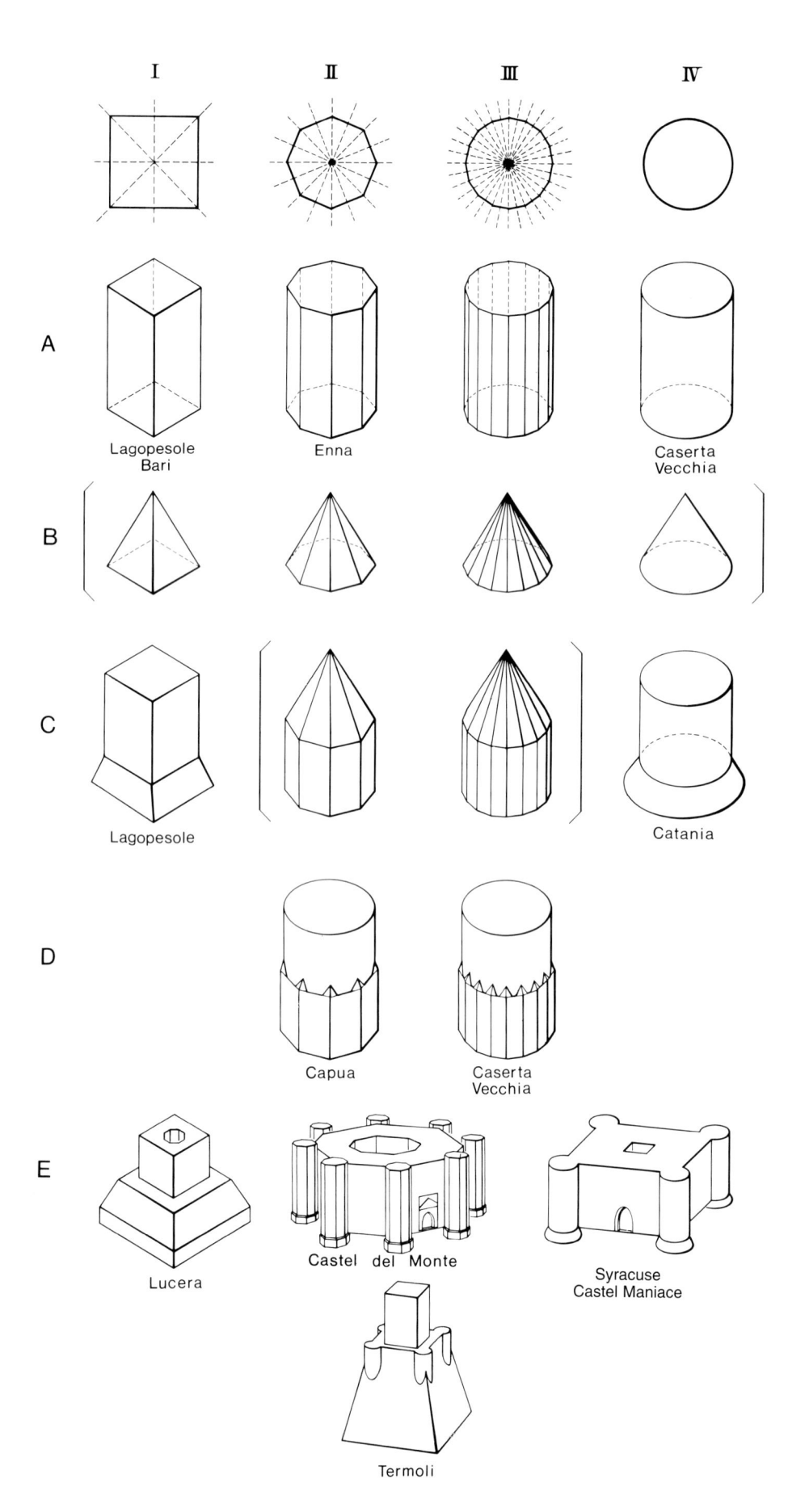

	I	II	III	IV
A	Lagopesole Bari	Enna		Caserta Vecchia
B				
C	Lagopesole			Catania
D		Capua	Caserta Vecchia	
E	Lucera	Castel del Monte	Syracuse Castel Maniace	
		Termoli		

155 Systematic comparison of the stereometric figures and forms frequently occurring in Hohenstaufen architecture between 1230 and 1250. The chronological sequence is not followed here.

In the top row appear the basic geometric figures with their lines of symmetry.

Row A shows the prisms or cylinders developed from these two-dimensional basic figures, row B the corresponding pyramids or cones. The figures in row B do not occur by themselves and are therefore shown within brackets.

Combinations of A and B do occur, however: C I and C IV.

The combinations C II and C III again do not occur by themselves but in connection with A IV: D II and D III.

Row E presents combinations that are found in the masterpieces built between 1230 and 1250. These are exclusively combinations of A, [B], and C.

Above and beyond this syntax of forms, it is of interest that the initial row of square–octagon–decahexagon–circle conforms in its turn to the sequence that symbolizes the transcendence from the earthly to the heavenly, which is most evident in the Islamic mausoleums, but also in Christian sacred buildings.

III

The Language of Forms in Hohenstaufen Architecture

Having analyzed the essential characteristics of the Hohenstaufen buildings in southern Italy erected during the decisive period of development covering the third and fourth decades of the thirteenth century, let us turn now to a systematic classification of these morphological 'shapes'. To begin with, it should be noted that two fundamental types served as the inspiration: the Norman donjon and the square citadel. The latter had its origin, as we have seen, in the Roman *castrum*, whose ground plan from the *limes arabicus* reached the area of southern Italy and Sicily via the Arab fortresses.

What are the characteristic forms of Hohenstaufen architecture in southern Italy with which we have become familiar in the preceding survey? Figure 155 is an attempt to classify them. In the top row is the *square*, from which the *octagon*, the *decahexagon*, and finally the *circle* are developed — the typical ground-plan shapes of the buildings discussed here. In the second row we find the respective *prisms* and the regular *cylinder* that stand on these outlines. We have encountered all of these stereometric bodies in the buildings. In the third row are the corresponding *pyramid* forms, including the *cone* (B); these forms do not occur independently, but only in connection with the correlative prisms or the cylinder (see row C). Some of the highly developed buildings of the later period unite all of the stereometric elements shown — but only and exclusively these.

It appears to be of prime importance that this catalogue of forms is completely self-contained and includes only certain connected groups of stereometric elements. Aside from the fact that we are thus far removed from the Gothic catalogue of geometric figures,[114] which contains the triangle and the pentagon (Chartres) as well as the hexagon and the non-square rectangle, the impressive element here is the sophisticated design concept of the 'classical' period of Hohenstaufen buildings, which moves exclusively between the square and the circle with their geometric and stereometric derivatives. This realization is underscored by the observation that secondary elements such as column bases and capitals (Figs. 156 and 157) or the circumferences of spiral staircases and stairwells also move within the framework of this hierarchy of forms: octagonal column bases, eight-part domed vaults in Castel Ursino and Castel del Monte (p. 48, Fig. 52 and p. 102, Fig. 147), flights of sixteen steps. The triangle or the hexagon occurs only exceptionally, for instance as a vault over cylindrical stairways in corner towers of Castel del Monte[115] and as a 12-step newel in Enna. In cylindrical rooms such as tower 7 of Castel del Monte (p. 102, Fig. 148) it is geometrically easier to construct the hexagon than the octagon from the radius of the circumscribed circle.

Thus, it appears that for the designs of all buildings uniform laws pertaining to a geometric system were passed down, to which the free creative will of the individual architects was subject. Whether there was a central design authority directly subordinate to the emperor for the buildings erected after the crusade is open to speculation. It becomes clear, moreover, that a creative precept obtained which was independent of central European Gothic.

Let us summarize the morphological features of the monuments considered up to this point:
1. Clear regularity of the outlines, based on the square or the circle
2. Manifold symmetrical relationships of the centrally planned buildings
3. Building substances with crystal-like structure arranged in stereometric elements
4. Plastic consistency of the building substances; superficial linear ornamentation is ignored — in contrast to Islamic architecture — and substituted by plastic decorative elements with a structure that stems 'from within'

In order to grasp the 'plastic consistency' of Hohenstaufen architecture, it is helpful to define more clearly the *nature* of plasticity. What is this plasticity that we see carried to perfection in Greek art — in sculpture as well as in archi-

156 Castel del Monte, Apulia: Hall IV on the upper story. Capital with eight-sided cover, of which five sides are visible. The thickness of the incrustation can be deduced here as well, from the marble rudiments on the wall.

tecture? Definitions have been proposed again and again to explain this phenomenon using a set of abstract contrary terms. With his theory of the opposite nature of the tactile and the optical, i.e., the tangible and the visible in the fine arts, Alois Riegl took an important step toward an understanding of the nature of sculpture, its creation and its effect. In his day he had to oppose a theory of aesthetics which viewed the essence of sculpture as founded mainly in its visibility from all sides. Riegl's theory was then taken up by Heinrich Wölfflin in his *Grundbegriffe* (Basic Concepts), which appeared in 1915 and became generally understood as the antithesis of the optical and the haptic. The two antitheses of Alois Riegl and Heinrich Wölfflin state that we are able to apprehend optical, 'pictorial' impressions with our eyes, but that we can grasp the truly plastic, three-dimensional perceptions through our sense of touch, with our eyes closed.

Prior to Alois Riegl and Heinrich Wölfflin, Johann Gottfried Herder had attempted to put the nature of plasticity into words, and he did so in such graphic, enthusiastic language that Herder's work *Plastik* fascinates the reader even today.[116]

The component of movement is also required, however, for a full grasp of the three-dimensional character or the plasticity of architecture. The plastic art of a building can be completely comprehended only by walking around and through it. The possibility of identifying by touch offered by three-dimensional objects is replaced in the case of architecture by the movement of the observer and the accompanying optical shift of the building elements in space. The road that leads up to Castel del Monte describes a circle around the building, thus cultivating the impression of three-dimensionality and plasticity in the manner discussed.

The stereometric-plastic nature of Hohenstaufen buildings allows of hardly any superficial, two-dimensional decorative elements. Wherever practical necessities constrain to linear solutions — for instance, with windows and portals (p. 82, Fig. 116 and p. 37, Fig. 30) — three-dimensional elements immediately come into play: through columns and pilasters, portals and windows acquire a deeply graduated solidity.

Hermann Weyl was of the opinion that art had never been concerned with spatial ornaments; these were to be

157 Castel del Monte, Apulia: Hall VII on the upper story. Eight-sided column base with three-column bundle of light-gray figured marble.

The Language of Forms in Hohenstaufen Architecture   111

found in nature, however, namely in the form of the arrangement of atoms in crystals, the so-called Laue diagrams.[117] Hermann Weyl was not familiar with the three-dimensional ornamentation of Hohenstaufen architecture. In our discussion of the individual Hohenstaufen buildings it has repeatedly been pointed out that one of their characteristics is the almost complete lack of linear, two-dimensional ornamentation. In its place there is an enjoyment of the surfaces and edges of the stereometric form, which are always perceived as the boundaries of physical masses. How strong an effect the sense of plasticity has is shown by the few existing ornaments, which of course are of three-dimensional nature — be it that building materials of different colors have been chosen for different stereometric elements (e.g., Caserta Vecchia and Capua), or that the plastic-decorative effect of intersecting elements has been used.

With this sense of three-dimensionality, Hohenstaufen architecture is closest to the art and architecture of the Greeks. This makes particularly understandable Frederick II's fondness for classical sculpture, with which he surrounded himself and which he spared no effort to display in his castles. He recommended that the school of sculpture he had founded orient itself according to classical models, and one is surprised and impressed even today by the truly classical character of the surviving sculptures in the museum of Capua, and of the architectural sculpture of Castel Maniace and Castel del Monte. Presumably, this school of art provided an important impetus for a new sense of plasticity in the early Renaissance — consider, for instance, Nicola Pisano (ca. 1225–1278), who possibly came from an Apulian school of sculpture.[118]

IV

Castel del Monte: Design and Construction

The geometric-stereometric structure of Castel del Monte and its ground plan described here invite a mathematical definition, which has not previously been attempted. We see a monohedron group with 16 elements: eight reflecting planes and eight rotations; it is an 'automorphic group'. The multiplicity of the symmetrical relationships is increased by the 'homeomorphic' connections, i.e., the 'similarities' between the large octagons of the main building and the octagon of the courtyard, and further with the eight octagonal towers arranged within the same system of axes. Not only to the mathematician but also to the historian of art and architecture, it is of great scientific interest to inquire into the nature of these symmetrical relationships — the more so, as we know that since antiquity the great architects and visual artists have been indebted to mathematics in terms of practical geometry.[119]

This is not surprising, as surveying and architecture, which require the exact measurement of horizontal and vertical projections, are the fathers of geometry.[120] We know that the builder of Hagia Sophia in Constantinople, Isidorus of Miletus, composed a supplementary chapter to Euclid's *Elements* concerning the angles of regular bodies.[121] Piero della Francesca (ca. 1415–1494) also wrote a treatise on regular bodies, and the mathematician Luca Pacioli, a friend of Leonardo, wrote a book about the golden section. Dürer examined and described the proportions of the human body and dealt with the regular divisions of planes, as did Daniele Barbaro, the patriarch of Aquileia, after him. The seventeenth-century Italian architect Guarino Guarini was originally professor of mathematics and philosophy in Messina.

By a stroke of luck, the lodge book (MS fragment 19093 in the National Library of Paris) of Villard de

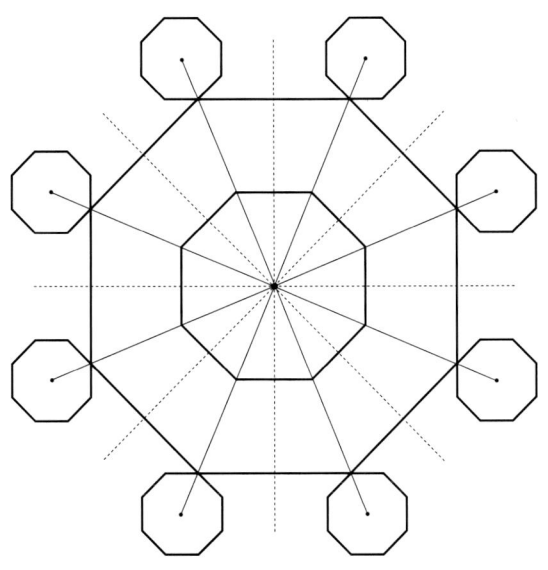

158 Castel del Monte, Apulia: Ground plan with lines of symmetry.

Honnecourt, a contemporary of Frederick II, has been preserved; a complete critical edition was published in German by Hans R. Hahnloser.[122] It gives us an excellent idea of the knowledge and manual skill that an architect of that period in Europe possessed. The knowledge imparted by Villard is supplemented by records of late Gothic architects of the fifteenth century in the practical guidelines of Mathes Roriczer and Hanns Schmuttermayer.[123]

In Castel del Monte we perceive the dominant role of mathematically definable symmetrical relationships and must therefore examine their significance. The word 'symmetry' comes from the Greek *symmetros* and is best translated as 'harmony'. This implies much, much more than simply 'mirror imagery', which it generally means in current usage and which is only a particular case of the general concept of 'symmetry'. 'Harmony' is associated with beauty. Those dimensions are symmetrical, i.e., harmonious, that are in correct proportion to one another. Such a harmonious-symmetrical system of relationships, for example, was the *Canon*[124] of the Greek sculptor Polycleitus from the fifth century B.C. His teachings about representing the harmony of the human body and about the relationship of the individual parts to one another and to the three-dimensional figure as a whole constituted the height of Greek art theory.

It is not surprising that symmetry — which we have here equated with harmony and beauty — is employed repeatedly in art as a stylistic device. Here we touch on a question of philosophical aesthetics and epistemology which we can mention only briefly, namely whether the artist discovered the aesthetic quality of symmetry in nature and imitated it, or whether the aesthetic value of symmetry has common roots in nature and in art. Plato assumed that the mathematical idea was the common origin of both: The mathematical laws that govern nature are the source of symmetry in nature; its source in art is the intuitive grasp of the idea in the mind of the *creative* human being.

We have attempted to define the symmetrical relationships that distinguish all Hohenstaufen buildings, and to an especially great degree Castel del Monte; at the same time we have seen that they characterize the nature of the building to the full and make it comprehensible. This is an observation that excludes from all further comparisons all those buildings that, while they may display distant similarities in the layout of rooms, lack a clear system of symmetrical relationships and thus belong to a different morphological concept, to another 'form'.[125] This raises questions of methodology relating to art history. The common method of comparing individual forms and motifs covers only one dimension of a work of art. The results are relatively good when homogeneous schools of art are considered — central European, French, etc. — where the individual form is presented and further developed in parallel with the whole. Here the inference from the single motif to the development of the whole is permissible. In the case of more complex structures, particularly when they appear and evolve within schools and regions of art that overlap developmentally and temporally, this inference from the individual to the whole is no longer tenable. Critical viewing and comparisons of the overall morphological concept, the 'form', lead to more reliable results.

1 *The Octagon*

In analyzing the language of forms in Hohenstaufen buildings we have discovered the octagon, in addition to the square and the circle, as a predominant basic form; we have encountered the hexagon, however, only in connection with the circle, with whose radius the sides of the hexagon correspond. The outline and symmetries of the construction plan of Castel del Monte are related almost exclusively to the octagon. Therefore, we must look into the significance of this geometric figure.[126]

In classical Greek geometry, lines and the circle were the basis of all considerations and reflections. The circle is easy to construct, and the inscribed equilateral hexagon and thus the equilateral triangle are practically self-evident results of its radius; all of these are starting points for more extensive geometric operations. In addition, the square has fascinated mathematicians, philosophers, and artists from antiquity up to modern times. It is no coinci-

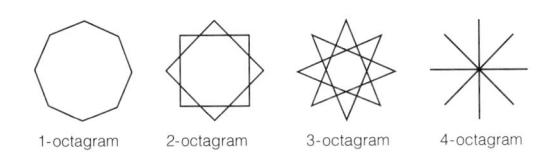

159 Polygons and star polygons based on the number eight (from F. L. Bauer).

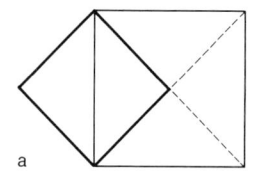

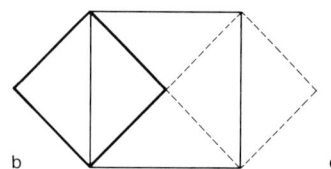

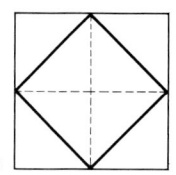

160 Bisection of the area of a square. Figures a, b, and c are taken from the lodge book of Villard de Honnecourt. Figure a corresponds at the same time to the representation in Plato's dialogue *Menon*.

dence that Leonardo da Vinci inscribes his figure of human proportions symbolically within the circle and the square.[127]

The square interested mathematicians and philosophers because of the impossibility, discovered in the period of Greek preclassicism, of finding a rational numerical proportion between the length of the diagonal and the length of the side of the square. The irrationality and incommensurability of the diagonal categorically contradicted the Pythagorean view that every dimension must be definable by a natural number.[128] A virtual crisis arose over this in mathematics, which was finally able to solve the problem only geometrically.

Plato was fascinated by the problem. In his dialogue *Menon* he describes how the area of a square can be geometrically doubled or halved by means of its diagonals, without having to use irrational or incommensurable numbers, in this case the quantity $\sqrt{2}$. With the Neoplatonists and Neopythagoreans, who had a strong influence on the philosophy of Islam, irrational quantities acquired a symbolic meaning.[129]

'Crossed' or 'revolving' squares are attributed to the Aristotelian theory of elements (*tetrasomia*), in which the forms of appearance of all elements are expressed in two pairs of opposites: hot – cold, moist – dry, which are symbolized by the four corners of a square. Corresponding to them is a second square with the four elements (Fig. 161) that represent the Platonic cycle of fire → air → water →

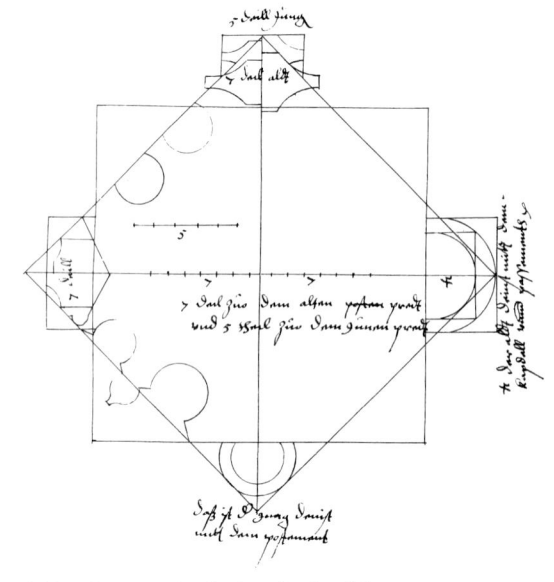

163 Geometric design for building components using intersecting squares. From Lechler's *Unterweisung* (Cologne, Historical Archive, manuscript Wf. 276, fol. 42r).

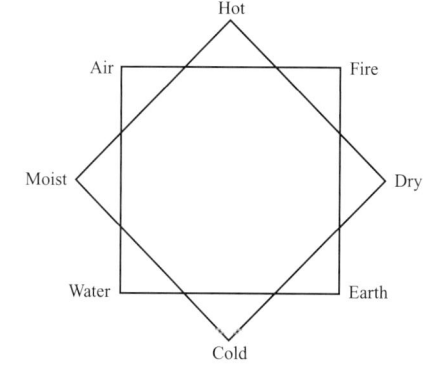

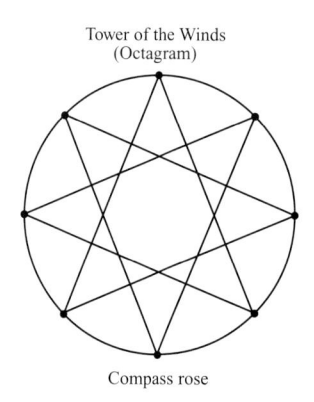

161 Illustration of Aristotle's theory of elements (*tetrasomia*). Two intersecting squares: one represents the mode of appearance of the elements (hot-cold, moist-dry); the other shows the four elements of the Platonic cycle.

162 Athens: Octagram of the Tower of the Winds, built by Andronicus Cyrrhestes (ca. 75 B.C.)

116 *Design and Construction*

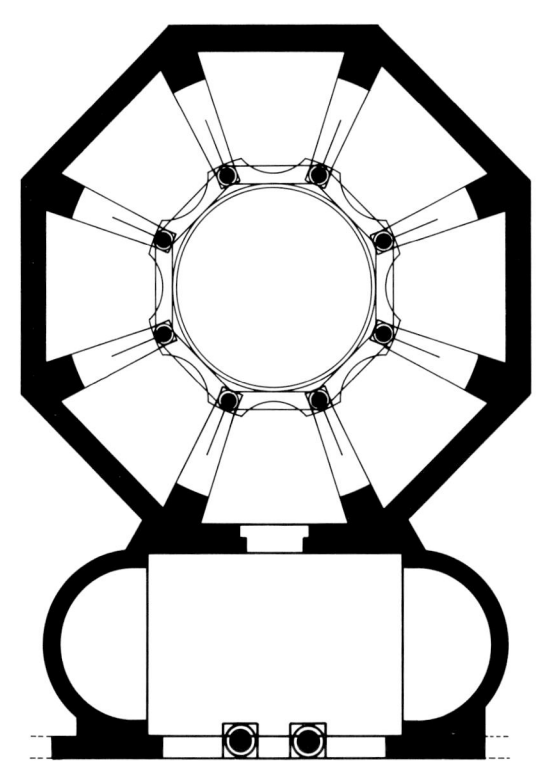

164 Rome: Basilica of St. John Lateran, baptistry (432–440).

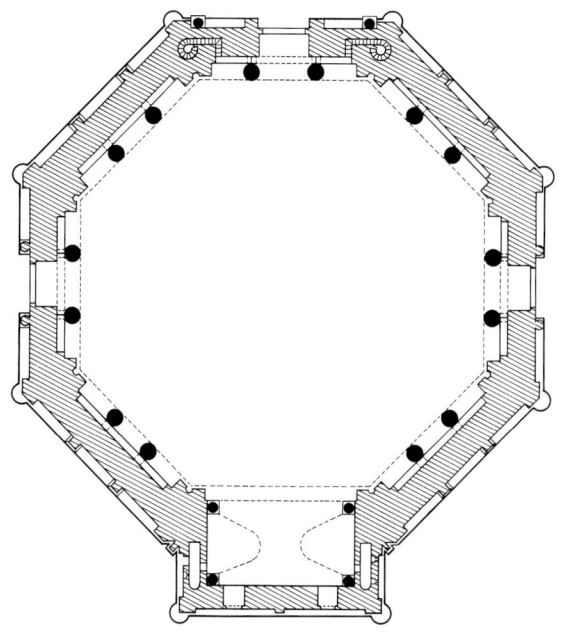

165 Florence: Baptistry of San Giovanni (1060–1150), plan of the ground floor.

earth → fire. It is assumed, however, that the symbol of the 'revolving' square itself goes back to older traditions.[130]

The Tower of the Winds in Athens, built by Andronicus Cyrrhestes (ca. 75 B.C.), is the architectonic representation of an octagonal symbolism (Fig. 162),[131] expanded and connected with the twelve signs of the zodiac.

We meet the figure of the 'revolving' square in art and architectural design as a type of 'guiding figure' as far away as China and Japan. It can also be found as a basic structural figure in late Gothic central Europe (Fig. 163). The great importance of the octagon and the eight-pointed star in the Christian as well as the Islamic hope of salvation and the closely related religious art derives from esoteric symbolism. It is the same symbology as that which underlies the octagonal form of late classical and medieval baptistries (Figs. 164 and 165).[132]

An inscription in the baptistry of the church of St. Thecla in Milan stems from St. Ambrose (339–397 A.D.); the first lines read as follows:

"With eight niches the temple rises to holy use.
The fountain is octagonal, worthy of the [sacred] gift.
The house of holy Baptism had to arise in the [sacred] number eight."[133]

If the square symbolizes the earth, the circle — in the absolute perfection of its form, the infinite number of its symmetrical axes, and its 'isoperimetry' — represents the immensity of Heaven. The impression of the horizon surrounding us and the vault of Heaven above us conforms with this symbology.[134]

From the kingdom of the earth, the square, into the infinity of Heaven and immortality, symbolized by the circle of the cupola — in many Islamic and Christian churches the octagon is the intermediate stage, occasionally supplemented by a decahexagon,[135] that leads to the circular tambour above which the cupola arches. In mosques and mausoleums, but also in the crossings of Christian churches overarched by cupolas, the transcendence to divine redemption, to paradise, and to immortality is symbolized in this way.

The symbolic meaning of the square and the circle applies equally in eastern Asia. In China as well, the square represents the earth, the four points of the compass, and the four elements; the circle, on the other hand, represents Heaven,[137] as evidenced by the round Temple of

The Octagon 117

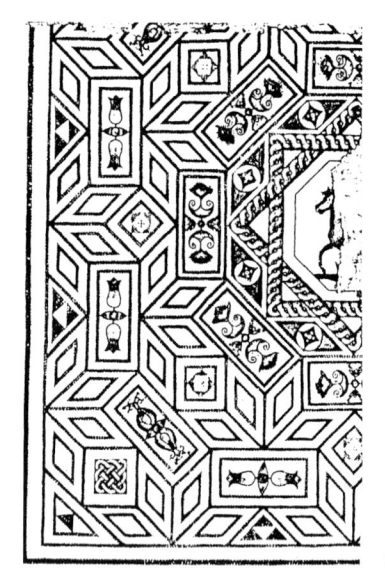

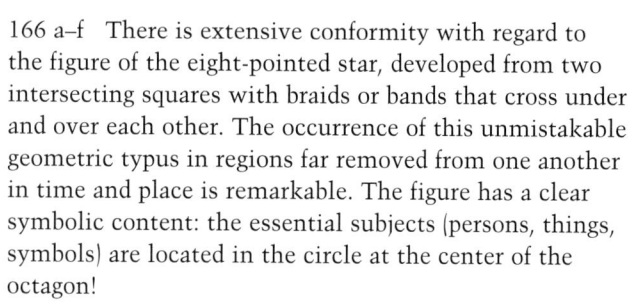

166 a–f There is extensive conformity with regard to the figure of the eight-pointed star, developed from two intersecting squares with braids or bands that cross under and over each other. The occurrence of this unmistakable geometric typus in regions far removed from one another in time and place is remarkable. The figure has a clear symbolic content: the essential subjects (persons, things, symbols) are located in the circle at the center of the octagon!

That it is found in Vienne, Isère, and in the Villa of the Piazza Armerina (Sicily) confirms its existence in the late Roman period, which includes, however, not only the western Roman but also the eastern Roman-Byzantine empire. Its appearance in China is surely due to the silk route, especially as the ornament also is frequently found in central Asia (Buhārā). It would be interesting to determine the source of this motif.

a) Vienne, Isère (France): Mosaic with an eight-pointed star pattern and a geometric background (from Lancha 1977; see also Artaud 1818), second century A.D.

b) Villa of Piazza Armerina, Sicily: Mosaic with the octagon developed from two crossed squares, beginning of fourth century A.D. Villa of Piazza Armerina, sacellum of the Lares, plate VI, Fig. 22.

c) Codex Dioscurides 'De Materia Medica' (prior to 512 A.D.). Vienna, Austrian National Library: Presentation miniature with Princess Anicia Juliana in an eight-pointed star. The illustration is evidence of the Greek-Hellenistic tradition of this eight-pointed star symbolism.[136]

d) Military banner of the Almohad caliph Abū Ya'qūb Yūsuf II, which fell into the hands of the Christians during one of the decisive battles of the *Reconquista*

118   *Design and Construction*

near Las Navas de Tolosa (1212); currently at Las Huelgas cloister, Burgos (from Dodds 1992). Here there are even two sets of two squares each that connect with one another to form two octagons, by means of under- and overcrossing of the sides — a smaller inner one and a more substantial outer one. In the latter case, the eight sides of the squares are extended to form a star octagon, thus forming the ground-plan figure of Castel del Monte (see p. 159, Fig. 226).

e) Beijing, Forbidden City: Taihe Dian ('Hall of Supreme Harmony'). The largest structure within the emperor's palace, on the three-stepped Dragon terrace, is also one of China's most magnificent wooden buildings (1669, restored several times). In the center of the interior rises the emperor's throne as the midpoint of the empire. In a square coffer of the wooden ceiling, directly above the throne, we see the octagon formed of two intersecting squares. In its middle, in a circular field, is the dragon, symbol of the emperor and the son of heaven, with pearls hanging beneath it, the image of purity. A more significant place for the octagon, the sign of the supernatural, is hardly conceivable in China.

f) Buhārā, Uzbekistan: modern metalwork with the eight-pointed star formed from two crossed squares. The under- and overcrossing of the band or the knotted braid is the same as seen in a, b, and c above. The metalwork shown here was produced in 1994; however, the motif is an ancient one in Buhārā and a kind of symbol of the city.

Examples a-f shown here substantiate the existence and great symbolic importance of the 2-octagram and the eight-pointed star developed from two crossed squares. It stems from the treasury of ornaments and symbols of Graeco-Roman times. It was adopted into the world of Islamic forms and reached China (via the silk route?) as a motif laden with symbolism.

The Octagon 119

167 Nara, Japan:
Horyu-ji. Octagonal 'Hall of Dreams', Yumedono (739 A.D.). Founded by Prince Shotoku (573–621), the Horyu-ji is the oldest preserved temple area in Japan and one of the 'Seven Great Temples of Nara'.

Heaven in Beijing. Buildings with an octagonal ground plan are numerous in China and Japan and are often connected with prominent figures of religious life. The octagonal Hall of Dreams of Prince Shotoku (573–627 A.D.) in the Horyu-ji temple at Nara (Fig. 167) is a particularly beautiful example of this. No less impressive is the octagonal *Keiguin*, dating back to 1251 A.D. (Fig. 168), which stands in a separate area of the northeast corner of the Koryu-ji in Kyoto. In each case the construction is set off from the complex of buildings surrounding it because of its special purpose. In the Horyu-ji temple it was the devotional hall of Prince Shotoku; the *Keiguin* contained

168 Kyoto, Japan:
Koryu-ji. Octagonal *Keiguin*, which contained a likeness of Prince Shotoku showing him at the age of 16 years.

120   *Design and Construction*

a likeness of the prince, showing him at the age of 16, as well as a statue of the Nyoirin-Kannon and a picture of the Amida-Nyorai.

In the most important hall of the imperial palace in Beijing, the Taihe Dian (the 'Hall of Supreme Harmony', 1669, which has been restored several times), at a preferred place on the ceiling above the raised throne of the emperor, we find the familiar motif of the intersecting squares that enclose a circle with the dragon, symbol of the son of Heaven (Fig. 166e). The dragon is holding one large and six small pearls, symbols of purity.[138] A corresponding coffer with intersecting squares over the imperial throne is found in the 'Hall of Cosmic Union' (Jiaotai Dian).[139] In the 'Hall of Mental Cultivation' (Yangxin Dian) the crossed squares are no longer visible, only the resultant eight-pointed star.[140] This form recurs in the 'Hall of Abstinence' (Zhai Gong), also within the complex of the imperial palace.[141]

In the cases described these coffer compositions are found above the seat of the ruler; they can also be seen at the heads of important Buddha statues. The form and symbolism of these coffers can be traced back to the Jin dynasty — for instance, to the hall of the Jingtu temple (1124 A.D.) in Yingxian, Shanxi province. There the coffers have only two levels: the square and the octagon with the dragon. A further example from an earlier period is offered by the Shanhua temple in Datong, also in Shanxi province. Over the statue of Buddha in this temple hangs the square, from which the recessed octagon is carved, with a further-recessed circle in which two dragons lie. The temple dates originally from the Tang period and was restored following a fire in 1128 A.D.[142]

Thus, in the region of eastern Asia as well, the octagon is endowed with numinous power and serves as a logogram for the overall picture of the world.[143] Those who think it is going too far to trace the symbolism of the octagon from Europe all the way to China should take into consideration the fact that old and close contacts existed between Persia, India, and China, especially during the Tang dynasty. The spread of Buddhism is characteristic of these connecting footpaths of cultural history, just as it is of the great old trade routes.

Returning to the western part of Asia, in the Abbasid capital of Baġdād we find an elaborate square Kufic

169 Baġdād, Iraq: Square-octagonal-shaped calligraphy in Kufic script with the names of Allāh, Muḥammad, and the ten prophets. Building ornament from the twelfth/thirteenth century.

The Octagon 121

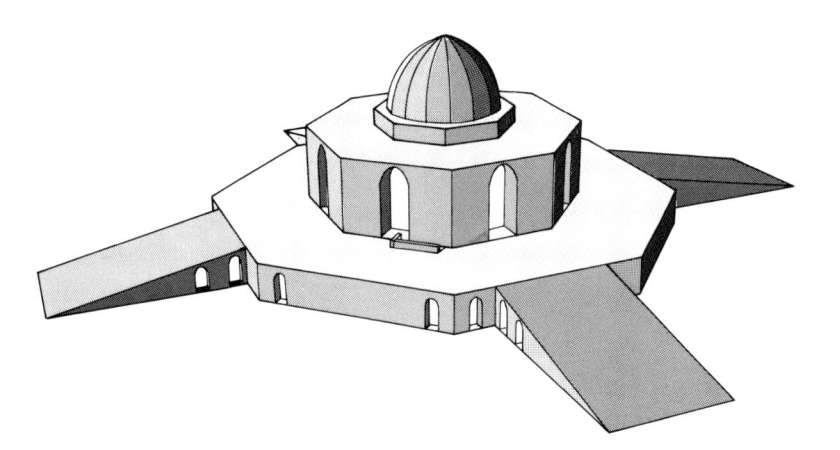

170 Sāmarrā, Iraq: Reconstruction of the original form of the mausoleum of Qubbat aṣ-Ṣulaibīya (after Northedge 1987). Pictured is the octagonal podium with four ramps on which the building rises.

inscription from the twelfth or thirteenth century giving the names of Allāh, Muḥammad, and the ten prophets (Fig. 169). It serves to decorate a building; its outer form is that of a large square, against which an octagon stands out. Its numinous character is obvious.[144]

Particularly illuminating in our connection is the astonishing abundance of square or octagonal buildings in the Islamic region, most of which are mausoleums: the earliest example stands on the west bank of the Tigris in Sāmarrā, the Qubbat aṣ-Ṣulaibīya (Figs. 170 and 171). It is not certain whether the building was originally planned as a mausoleum.[145] It is an octagonal central construction with an octagonal ambulatory on an octagonal podium. In the interior is a square room that changes into an octagon above, upon which the cupola rests.[146]

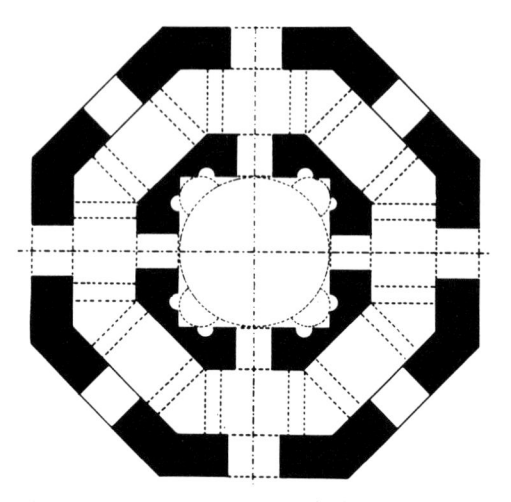

171 Sāmarrā, Iraq: Ground plan of the mausoleum of Qubbat aṣ-Ṣulaibīya (862 A.D.).

This type of mausoleum continued to be built in the Islamic world of central Asia. An excellent example is the mausoleum of the Samanids in Buhārā.[147] It was erected as a family grave by the Samanid Ismāʿīl between 892 and 907 A.D. A careful description and analysis of its dimensions and their proportions was made by M. S. Bulatov.[148] This mausoleum with a square ground plan is cube-shaped. The square gives way to an octagon and then to a decahexagon, on which the cupola stands (Fig. 172).

An old trade route led from Buhārā to Marw; the latter was captured by Abū Muslim for the Abbasids in 747 A.D. and became a center of Islam. Numerous mausoleums are found there with square foundations or even in the shape of a cube, i.e., with a total of six square sides. The square is carried over into an octagon, above which rise the circular tambour and the cupola. The mausoleum of Sulṭān Sanǧar in Marw (1157 A.D.) displays the same shape:

122   *Design and Construction*

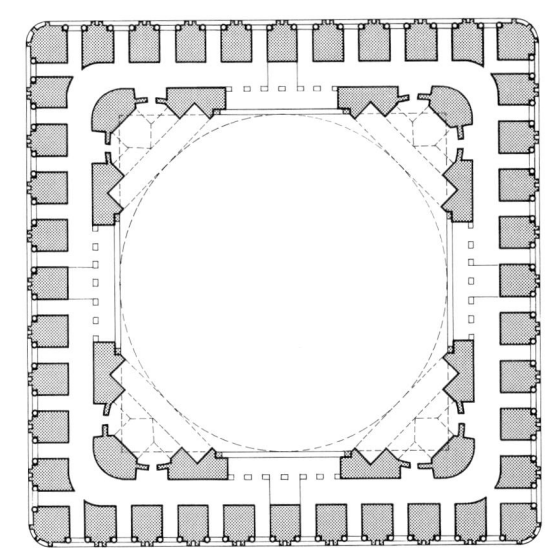

172 Buhārā, Uzbekistan: Ground plan of the mausoleum of the Samanid Ismāʻīl, around 907 (from L. I. Rempel).

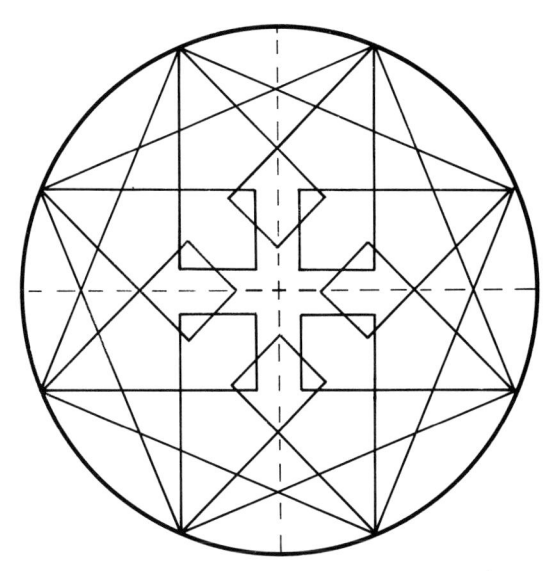

173 Marw, Turkmenistan: Mausoleum of Sulṭān Sanǧar (1157 A.D.), eight-pointed star motif in the cupola (from Bulatov 1988). The master builder was Muḥammad Atsiz.

Above a cubic foundation rises a high tambour that is cylindrical on the outside, octagonal on the inside, above it the cupola with the eight-pointed star motif (Fig. 173).[149]

The first octagonal mausoleum in the Islamic region of India was erected for H̱ān Ǧahān Tīlangānī in the Tuġluq period, 1368 – 1369 A.D. It served as the model for later octagonal mausoleums.[150]

An octagonal example from the early Ottoman period, ca. 1421, is the 'Green Tomb' in Bursa.[151] The eight-sided mausoleum of Šīr Šāh Sūr in Sasāram, built in approximately 1540, is especially splendidly appointed — with a platform in a lake.[152]

One of the best-elaborated examples of an octagonal mausoleum with an ambulatory is that of ʻĪsā H̱ān Niyāzī (1561), a general of Šīr Šāh Sūr (1540–1545), in Delhi. It stands not far from the later-built tomb of Humāyūn (Fig. 187). On its corners there are round, projecting towers.[153]

In the Punjab province of Pakistan stands a group of hitherto scarcely known octagonal mausoleums of a strikingly independent type. Their *ground plan* is eight-sided *with round towers on each corner*.[154] The oldest building of the group is the mausoleum of the mystic Šāh Rukn-i-ʻĀlam (i.e., 'Pillar of the World'; 1334–1340 A.D.) from the period preceding the rule of the Great Moguls. It is considered to be the building most typical of the Tuġluq epoch. The massive octagonal construction of bricks with blue and white glazed tiles displays Iranian and central Asian influences. Not only the basic octagonal shape is of interest to us; particularly noteworthy are the towers with circular ground plans that are attached to each corner and taper upwards.[155]

This huge construction (Fig. 174), the sepulchre of the holy Rukn-i-ʻĀlam, was erected by King Ġiyāt̠ ad-Dīn Tuġluq in the middle of a broad fortification, where a Hindu shrine (perhaps a temple to the sun?) with a colossal solid-gold statue of the god Aditya is said to have been earlier.[156] The ground plan is repeated in the smaller upper story, on which the cupola rests.[157]

In Multān there is a construction of the same type that is approximately 100 years younger: the mausoleum of ʻAlī Akbar (Fig. 176).[158] It does not possess the three-dimensional power of the Rukn-i-ʻĀlam. The portal is flanked by two corner towers that are not cylindrical or conical like all of the others, but rather polygonal, with

The Octagon 123

174 Multān, Pakistan: Mausoleum of the holy Rukn-i-'Ālam, built during the first third of the fourteenth century.

eight visible lateral faces (Fig. 177). The number eight returns! These are not geometrically regular octagons, however.

A further building that follows the same concept stands south of Multān in Uč: the tomb of Bībī Ǧāwandī dating from the fifteenth century (Fig. 175). It has an effect similar to that of the mausoleum of Rukn-i-'Ālam (cf. Fig. 174).[159] The wonderful Persian tiles are largely visible even today and are an indication of the importance of this building and its cultural connections.

The mausoleums in Multān and Uč belong to a typologically homogeneous group and contrast clearly with other mausoleums of the Muslim area. The prime example of this group, the tomb of Rukn-i-'Ālam, appears to be

175 Uč, Pakistan: Tomb of Bībī Ǧāwandī (1494).

124   *Design and Construction*

176 Multān, Pakistan:
Mausoleum of ʿAlī Akbar (fifteenth century).

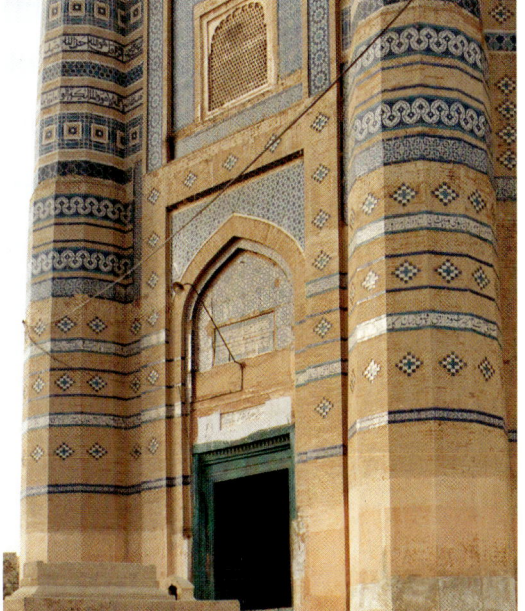

177 Multān, Pakistan: Mausoleum of ʿAlī Akbar (fifteenth century). Towers of the portal facade with eight visible sides.

already consummate, so that the question of forerunners is permissible but for the time being unanswerable.

Finally, attention should be called to a monument at quite another location: a large octagonal building in Ağdābīya on the Gulf of Sidra, i.e., far from the buildings mentioned up to this point, said to have been created by al-Qāʾim bi-amr Allāh Abū l-Qāsim (934–946 A.D.), the second ruler of the Fatimid dynasty.

The type of building that is octagonal from the foundation upward, which we have traced here through the centuries, appears to have been limited in the Islamic world primarily to mausoleums. Palaces and mosques generally have square plans. Above the base, however, this can change into an octagon, and finally into a circular tambour on which a cupola stands. An especially beautiful example is the Omayyad mosque in Córdoba. Two cupolas carried by powerful ribs rise above octagonal ground plans derived from the square. The plan of one

The Octagon   125

cupola — both were built between 961 and 966 — is based on the 'classical' figure of two squares 'rotating' toward each other by 45°. If we look at the cupola from below we see an interesting three-dimensional figure, with the corners of the squares resting on columns (Fig. 179) that mark the corners of the octagon. The corners of the square transfer, as it were, the weight of the cupola to the columns of the octagon. The interesting thing is that, strictly speaking, this cupola construction is conceived not as spherical, but rather as two-dimensional — in a flat geometric projection the two crossed squares are clearly discernible (Fig. 178).[160]

Thus we come across a peculiarity of Arabic architecture and the mentality on which it is based: It is not spatially or plastically, but rather two-dimensionally oriented, as evidenced in such varied ways by the splendid and ingenious Islamic ornamental art.[161] The appreciation of surface decoration achieves triumphs with artistically designed symmetrical figures that are based on mathematical considerations and constructed geometrically.

In principle, the same cupola construction as that in front of the mihrab of the mosque in Córdoba (Fig. 179) is repeated in the oratory of the Omayyad palace of Aljafería in Zaragoza (Fig. 180).[162]

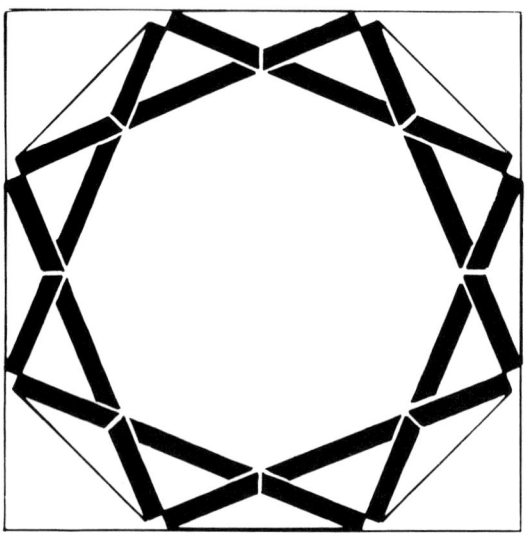

178 Córdoba, Spain: Ribs of a cupola in front of the mihrab of the Omayyad mosque (961–966 A.D.), plane projection.

179 Córdoba, Spain:
One of the two cupolas in front of the miḥrāb of the Omayyad mosque (961–966 A.D.). In a plane projection, the sturdy ribs form two intersecting squares turned toward one another by 45° (cf. Fig. 178).

126   *Design and Construction*

180 Zaragoza, Spain:
View into the cupola of the Omayyad Aljafería palace (second half of the eleventh century). Note particularly the distinct figure of the crossed squares in the center of the cupola.

181 Córdoba, Spain: Omayyad mosque. Schematic representation of the wall arches from Fig. 182, projected on a plane.

182 Córdoba, Spain:
One of the two cupolas in front of the miḥrāb of the Omayyad mosque (961–966 A.D.).
In a plane projection, the sturdy ribs form a perfect eight-pointed star (cf. Fig. 181).

The second cupola in Córdoba (Fig. 182) is possibly even more interesting. Here as well, one is impressed by the powerful ribs that support the vault. The two squares turned toward one another by 45° are distinguishable in the center. It is not the eight corners of the squares that rest on the corner pillars, however; rather, it is the extensions of eight times two square sides that intersect one another in an extended circle and form an eight-pointed star (Fig. 181).

The Octagon   127

183 Torres del Rio, Spain: Interior view of the cupola of Santo Sepolcro, dating from the late Romanesque period.

184 Mīhne, Turkmenistan: Painted cupola from the mausoleum of Abū Sa'īd (beginning of the fourteenth century).

A similar figure can be found in the cupola of the Bāb Mardūm mosque in Toledo (990 A.D.) — also with strong ribs, which rest here, however, not on the corners of the underlying octagon, but on the middle of the architraves that connect the corners of the octagon with one another.

This solution was also used for the domed vault of the late Romanesque centrally planned Church of the Holy Sepulchre (Fig. 183) in Torres del Rio. Here as well, the ribs rest not on the octagon corners but on free-standing consoles in the middle of each side. Additional ribs lead from the corners up to the spandrels of the eight-pointed star, transferring part of the vault load onto the corner columns.[163]

A somewhat ornate polychrome version of this cupola structure from Córdoba is seen in the mausoleum of Abū Sa'īd in Mīhne (Fig. 184), dating back to the beginning of the fourteenth century,[164] where the eight-pointed star is likewise developed from two diagonally placed squares and ornamentally elaborated. A related example can be found in Samarqand in the mausoleum of Šādī Mulk Āqā from the year 1371 (Fig. 185).[165]

185 Samarqand, Uzbekistan: Eight-pointed star motif in the cupola of the mausoleum of Šādī Mulk Āqa (1371).

128   *Design and Construction*

186 Granada, Spain: Alhambra, Court of the Lions, Sala de los Abencerrajes (fourteenth century). Muqarnas cupola in the shape of an eight-pointed star developed from intersecting squares (Córdoba type, cf. Figs. 178, 179).

A magnificent example of an octagonal cupola in sophisticated stalactite work (muqarnas) can be seen in the Court of the Lions complex of the Alhambra dating from the end of the Naṣrid rule (fourteenth century). It belongs to the 'crossed-squares' type (Fig. 186).

2 The Eight-pointed Star

The example of Córdoba can be used to show the development of the eight-pointed star from the octagon. This figure has retained its symbolic power in the Islamic region down to the present day and goes back to pre-Islamic times. The figure is called the Arabic eight-pointed star, occasionally also the seal of Solomon, although the latter term is more appropriate for the six-pointed star (or star of David).

As the origin of the figure is uncertain, we shall continue to call it simply the 'eight-pointed star'. It is com-

187 Delhi, India:
Total view of the mausoleum
of Humāyūn (1565).

parable to a mandala. This mandala-like character is clearly to be seen on the standard of Abū Yaʿqūb Yūsuf II (Fig. 166d). There as well, the eight-pointed star is derived from the 'rotating' squares[166] and symbolizes the perfect world aspired to by Islam.

The numerous examples of octagonal buildings in the Islamic world indicate the position held in this region by the octagon and the eight-pointed star. Octagonal buildings are much less common in Europe, where they can generally be traced back to Byzantine prototypes — San Vitale in Ravenna, for example, or the cathedral at Aachen. The tradition of the baptistries has already been discussed, as have the octagons above the crossings of churches and cathedrals; an especially interesting variation from the period after 1322 is given by the wooden octagon of the cathedral at Ely in Cambridgeshire.[167]

Let us close this series with a further example from the Islamic area of India. It provides more insight into the

188 Delhi, India:
Mausoleum of Humāyūn (1565).
Ornamental mosaic with the
eight-pointed star motif on
a raised platform beneath the
cupola and above the tomb.

130   *Design and Construction*

189 Āgra, India: Tāǧ Maḥall (1632–1652). Main hall above the tombs with octagonal enclosure of the sarcophagi. The floor mosaic at the front displays one of the oldest Islamic octagon motifs.

190 The oldest Islamic star motifs are based on regular polygons that are inscribed within a circle (a). The elimination of certain line segments produces the simple eight-pointed star (2-octagram) or the characteristic 'Islamic' eight-pointed star (3-octagram, b–d; after Lee 1987, p. 182 ff.).

importance and development of the eight-pointed star in connection with a recognizable significance. In the mausoleum of Humāyūn in Delhi (Fig. 187), erected by his widow in 1565 and conforming to the general style of central Asia, particularly that of Tamerlane in Samarqand,[168] the construction of the eight-pointed star motif from the crossed squares can be distinguished on a mosaic floor (Fig. 188). It is indicative of the significance of this motif that it covers a raised area situated directly under the cupola and over the main grave of the mausoleum. We recognize a corresponding construction in the Tāǧ Maḥall at Āgra, where there is an octagonal enclosure of the sarcophagus in the same position and a mosaic floor with an octagonal motif (Fig. 189).

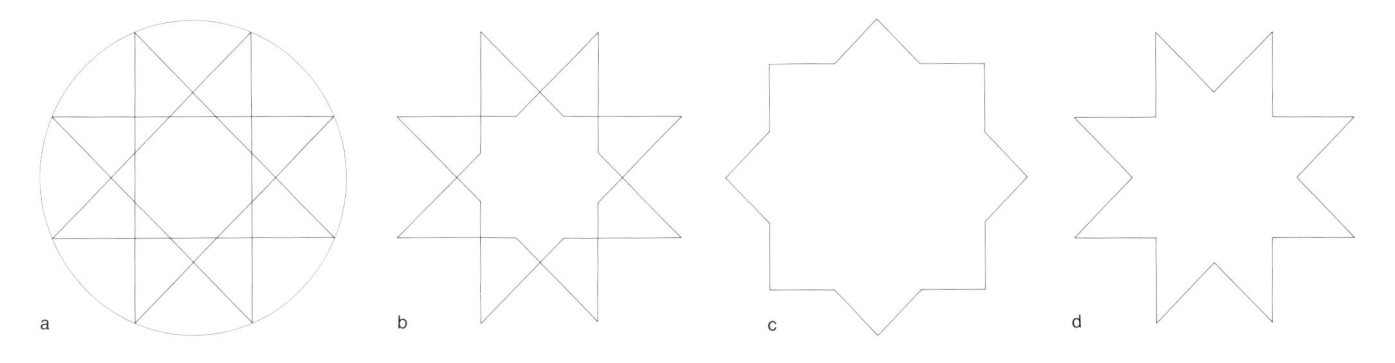

a b c d

The Eight-pointed Star 131

191 Plan for construction of the eight-pointed star from squares with an inscribed circle, still used today by Muslim craftsmen, e.g., in Morocco.

We are dealing here with the 'canonical' form of the eight-pointed star, which is derived from the 'rotating' squares and the circle (Fig. 190), as Issem El-Said and Ayşe Parman also descriptively elaborated for all Islamic ornamentation in their work *Geometric Concepts in Islamic Art*.[169]

French Islamists and historians of architecture have observed that contemporary Arab architects and artisans in Morocco apply age-old methods to create the eight-pointed star and other figures, making it possible for us to reconstruct them.[170] One result is the figure reproduced here (Fig. 191), which leads us back to the beginning of

192 Delhi, India: Mausoleum of Humāyūn (1565). My own graphic reconstruction of the original mosaic from Fig. 188. This drawing was done one year before I discovered the construction sketch shown in Fig. 191.

132   *Design and Construction*

193 The eight-pointed star in sketches by Leonardo da Vinci for centrally planned buildings. The example reproduced here is from the manuscript MS. B. (Paris) and is dated at 1498 or later.

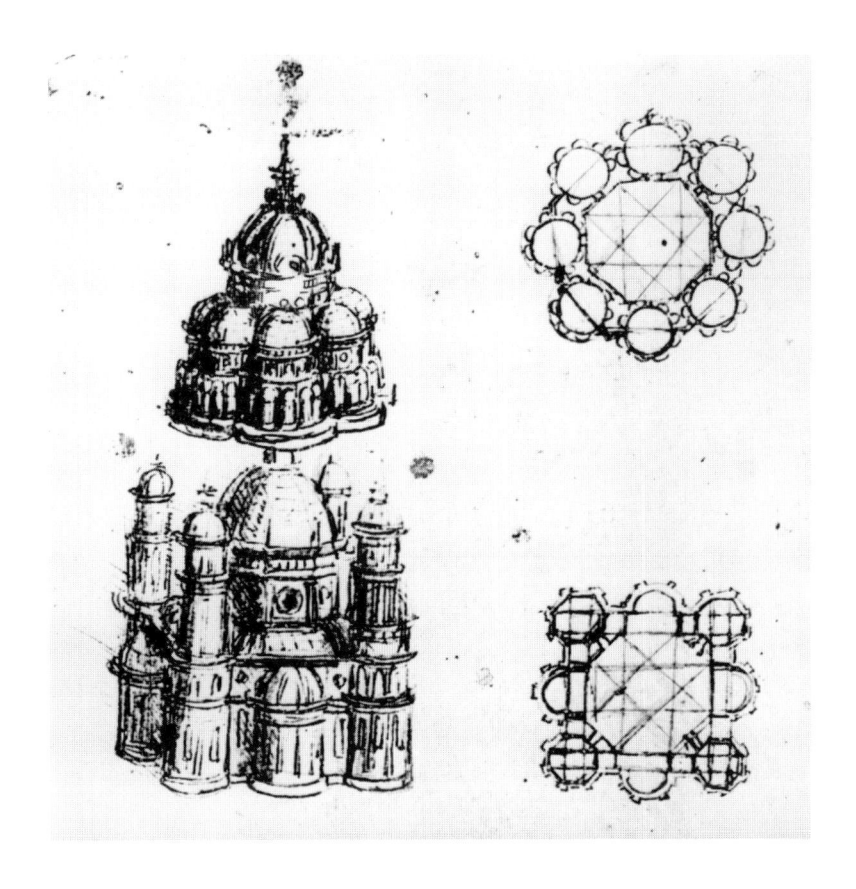

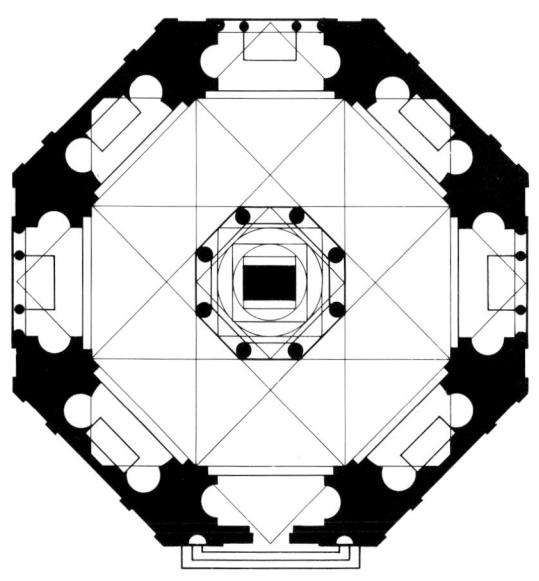

194 Sebastiano Serlio, design for an octagonal centrally planned building from the fifth book of his treatise on architecture, Paris, 1547.

our considerations. The workshop drawing of the motif made by present-day craftsmen is congruent with the mosaic from the mausoleum of Humāyūn in Delhi (Fig. 192). The Islamic draftsman starts from the square with its diagonals and diameters; within this square he inscribes the circle and connects each circle/diagonal and circle/diameter intersection to the two opposite intersections of circle/diagonal or circle/diameter; by extending the sides of the resulting two 'rotating' squares he then obtains the figure of the eight-pointed star.

It is particularly surprising to find the eight-pointed star again with Leonardo da Vinci. Although Leonardo did not distinguish himself as an architect, he and Bramante — like other artists of their time — were profoundly interested in the theory of centrally planned construction. There are many sketches by Leonardo concerning this subject (Fig. 193).[171] There is no doubt that in this and further sketches it was Leonardo's intention to outline the construction of the cupola itself.[172]

A purely octagonal centrally planned building was designed by Sebastiano Serlio in the fifth book of his treatise on architecture (Paris, 1547).[173] It is constructed exactly according to the principle of the two squares rotated

The Eight-pointed Star 133

195 Turin: View into the cupola and lantern of San Lorenzo (Guarino Guarini, 1668–1687).

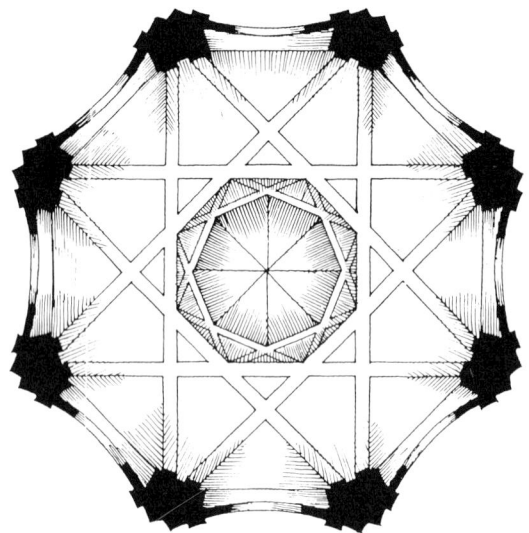

196 Turin: Cupola of San Lorenzo (Guarino Guarini, 1668–1687). Schematic drawing in which the larger, lower part of the cupola with the eight-pointed star motif stands out clearly from the smaller part above it with the motif of intersecting squares. With great ingenuity, Guarini set the two cupolas that stand next to each other in Córdoba on top of one another here (cf. Figs. 178 and 181).

by 45° with the resulting regular eight-pointed star. The *inner* octagon that is thereby generated represents here the *tribuna* resting on eight columns, the tabernacle above the high altar (Fig. 194).

It is possible that a direct line leads from these construction designs for centrally planned buildings to one of the most important and interesting Italian architects of the seventeenth century, Guarino Guarini (1624–1683; Fig. 195), who used the eight-pointed star as the supporting element in the cupola of the church of San Lorenzo in Turin.[174] A look at this cupola reminds us of the cupolas of Córdoba (Fig. 182).[175] Guarino Guarini was professor of *mathematics* and philosophy in Messina (1655) when he designed his first important buildings — here again, a close connection between architecture and mathematics! In Messina he is certain to have encountered Arab-Norman and Byzantine architecture. It is not known whether he saw the mosque of Córdoba himself, but the

134   *Design and Construction*

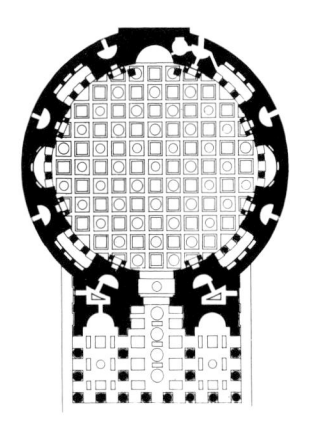

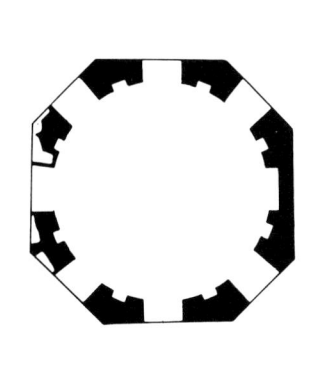

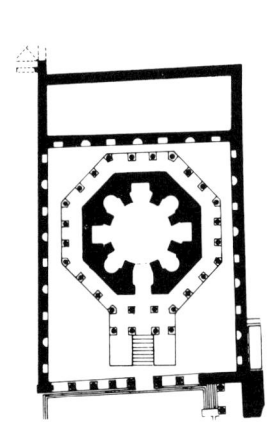

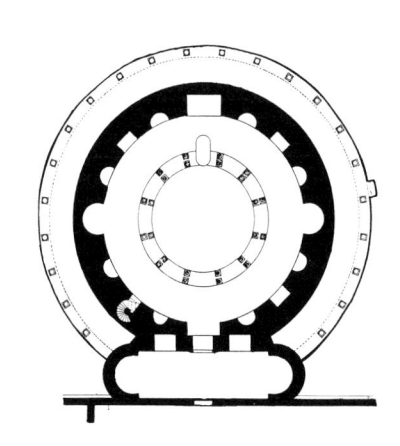

197 Rome: The Pantheon (118–128 A.D.).

198 Bajae, Campania: Temple of Diana in Roman thermae (second century A.D.).

199 Split, Dalmatia: Mausoleum (303 A.D.) in Diocletian's palace.

200 Rome: Santa Constanza (324–326 A.D.), mausoleum of Constantina.

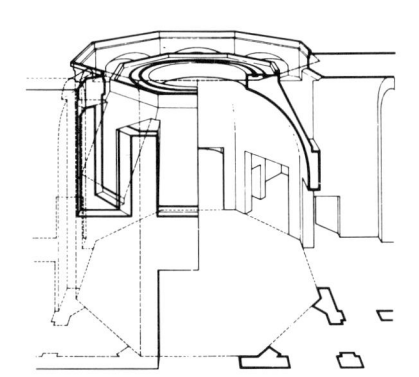

201 Rome: Octagonal structure in the Domus Aurea (64–68 A.D.), one of Nero's stately palaces.

interesting arrangement of *both* cupola structures from Córdoba *above one another* in San Lorenzo suggests that he did (Fig. 196).[176]

The appearance of the eight-pointed star motif in works by Leonardo da Vinci and Sebastiano Serlio raises the following two questions:

1. Was the motif of the eight-pointed star rediscovered?
2. Did this symbolically rich motif reach the area of Islam and the West via Byzantium and Syria?

There is much that speaks for the latter, but only a systematic examination of the history of centrally planned buildings in Europe and Asia Minor since Graeco-Roman times can bring a definitive clarification.[177]

The combination of central planning with the octagon can be found quite early, for instance in the Roman Pantheon (118–128 A.D.; Fig. 197), with an interior arrangement of eight niches. It may be inferred that Nero's Domus Aurea (64–68 A.D.; Fig. 201) contained an octagonal vaulted room, and the temple of Diana at Bajae (Fig. 198) also displays a fully formed octagon with a circular interior. The octagonal mausoleum in Diocletian's palace at Split (303 A.D.; Fig. 199) anticipates the ground plans of later Islamic mausoleums, such as that of Sāmarrā (Fig. 170). The interior of the circular mausoleum of Santa Constanza in Rome (324–326 A.D.; Fig. 200) has a sixteen-part arrangement. In San Vitale at Ravenna (540 A.D.; Fig. 202) not only the basic octagonal shape but also the eight-pointed star is clearly discernible as a leitmotif for the plan. The extensions of the eight square sides cross in the outer pillars of the octagonal building, i.e., at structurally essential points of load. With these eight-pointed star constructions a pair of opposite star points lies either on

The Eight-pointed Star   135

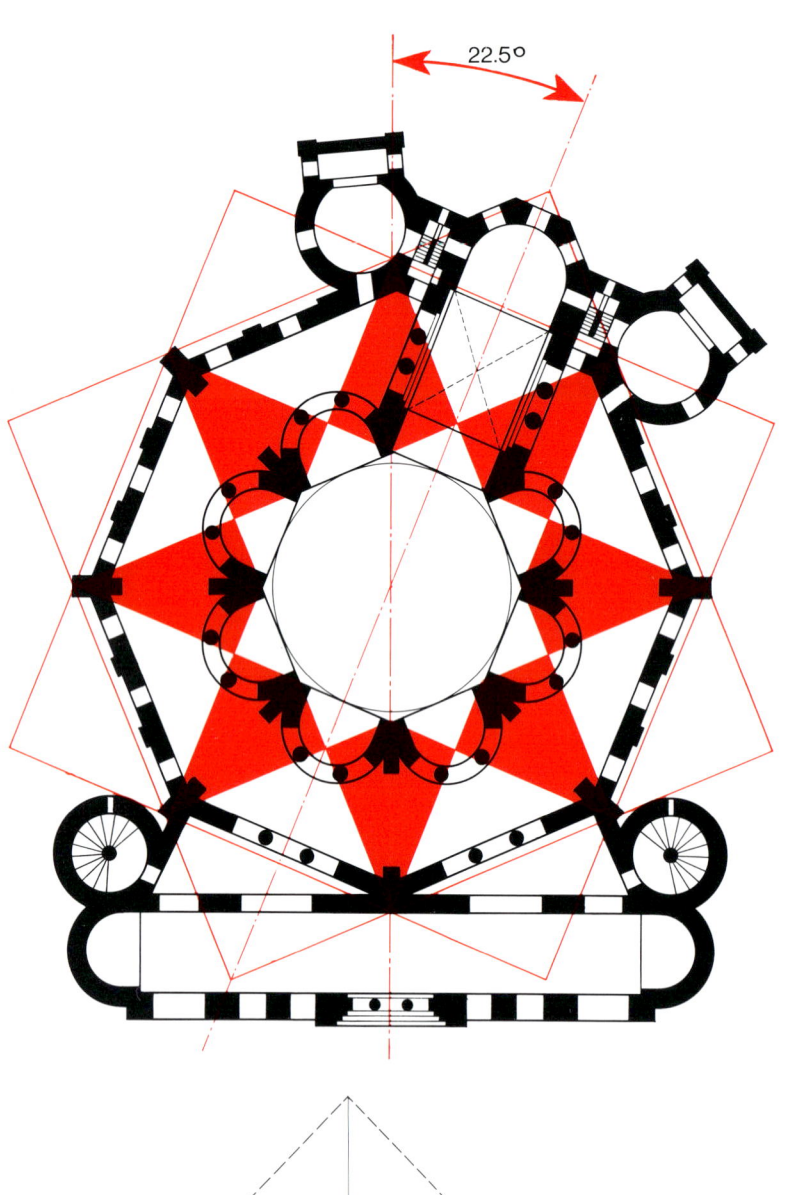

202   Ravenna:
Ground plan of the church of San Vitale (540 A.D.) with eight-pointed star drawn in.

203   Orthogonal and diagonal placement of the octagon.

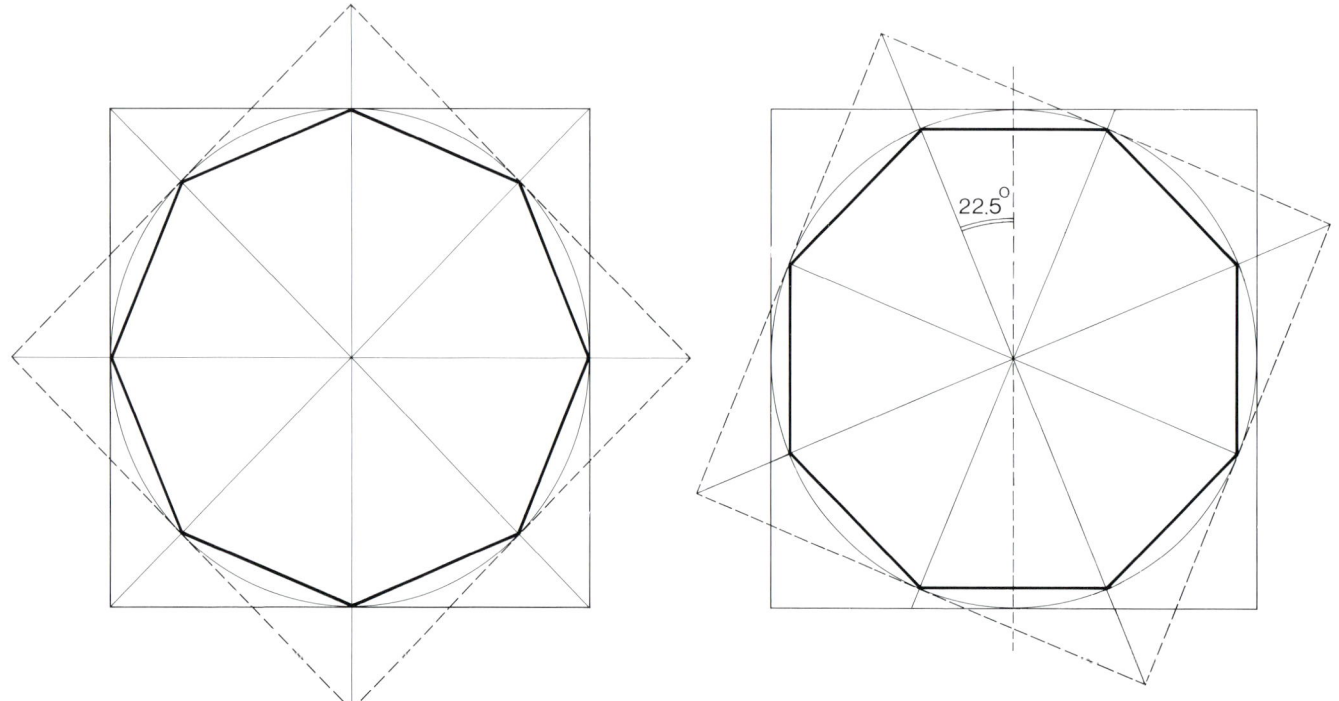

136   *Design and Construction*

204 Jerusalem: View of the Dome of the Rock (688–691 A.D.).

the building's main axis or, after rotation by 22.5°, on the diagonal of a square. Both axes are used in San Vitale, which explains the narthex rotated by 22.5° in relation to the main axis of the church (Figs. 202 and 203). The eight-pointed star as a constructional figure can also be found in the pilgrimage church of St. Simeon Stylites in Qal'at Sim'ān (490 A.D.), an octagonal centrally planned building.

The highlight of this series of buildings in the Byzantine architectural tradition is the Dome of the Rock in Jerusalem (Figs. 204–206), one of the most impressive monuments of the Islamic world.[178] Qubbat aṣ-Ṣahra was built by caliph 'Abd al-Malik between 688 and 691 on the rock upon which the great sanctuary of the Jewish faith, the temple of Solomon, had stood. In addition to this, it is the sacred rock on which Abraham prepared to sacrifice Isaac and from which Muḥammad took leave of the earth to enter heaven. Amid Christian-Byzantine surroundings, 'Abd al-Malik wanted to create a central sanctuary for Islam as meaningful as the Ka'ba in Mecca. It is conceivable that the nearby al-Aqṣā mosque (begun about 715) and the Dome of the Rock form a 'Constantine' group in the sense of basilica and anastasis, i.e., the Resurrection.

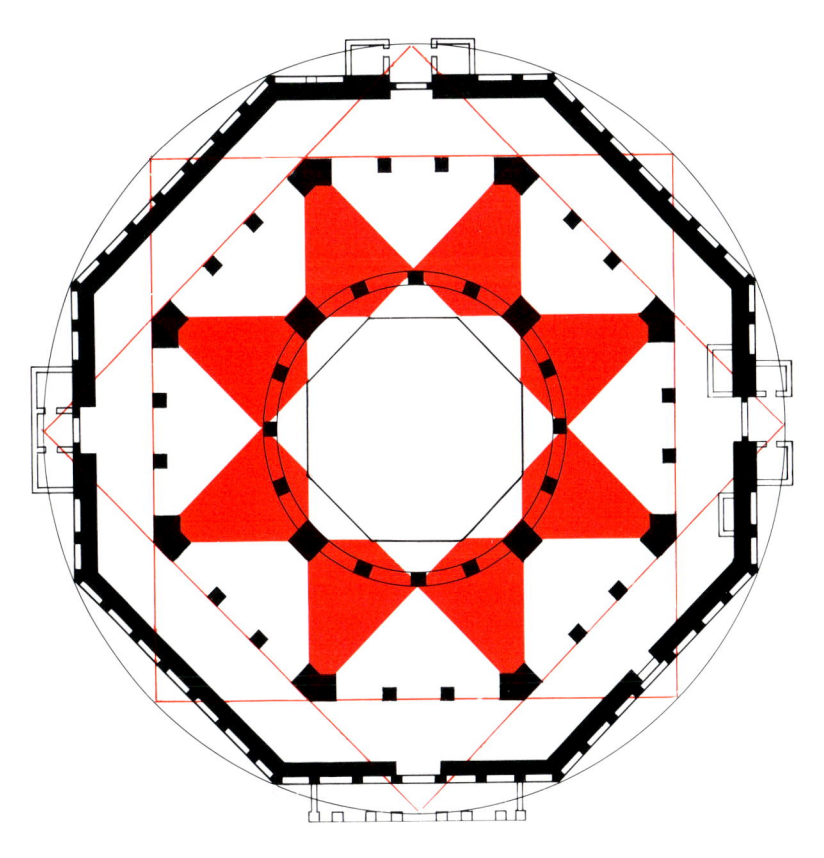

205 Jerusalem:
Ground plan of the Dome of the Rock (688–691 A.D.), with eight-pointed star drawn in.

The Eight-pointed Star 137

Many factors suggest this — such as the porch on the western side of the Dome, facing the mosque, more elaborately shaped than the other three portals, and the connections between the mosaic work on the walls in the interior of both sanctuaries.[179] It is the earliest Islamic shrine in the style of late Byzantine architecture, and its magnificence and harmonious proportions have been praised repeatedly, with justification.

We know that during his crusade in 1228/29 Frederick II visited the Dome of the Rock and that he had the construction of the cupola in particular explained to him.[180] It is highly probable that the ground-plan design was also explained on this occasion, as the emperor, known for his inquisitiveness, surely did not rest until he had a full grasp of the design principle of this widely renowned building; quite independent of its architectonic importance, its combination of Christian with Islamic tradition must have attracted his attention to an unusual degree. There is much evidence of his lively interest in architecture, which is clearly expressed in the originality and beauty of the architectonic designs for his citadels and castles.

It is not possible here to go into a detailed description of the Dome of the Rock. In connection with our observations up to this point, however, it may be said that the circle and the octagon play a dominating role. Beyond that, going back to an observation made by Mauss,[181] K. A. C. Creswell pointed out the various relationships and proportions between the circular centrally planned building, which supports the cupola, and the two ambulatories (Fig. 206). The four pillars of the center building, with three columns standing between every two, mark the corners of a square that is intersected by a further square rotated by 45°; the side extensions of this second square meet the eight pillars of the inner ambulatory. These eight pillars in turn stand at the intersections of two larger squares, which are also rotated toward one another by 45°. Oddly enough, Creswell appears not to have noticed the fact that we have here a fully formed *eight-pointed star*. The surprise is even greater when we become acquainted with the observations of the French architect Michel Ecochard,[182] according to which not only the proportions but also the absolute dimensions of the basic plans supposedly correspond with the ground plans

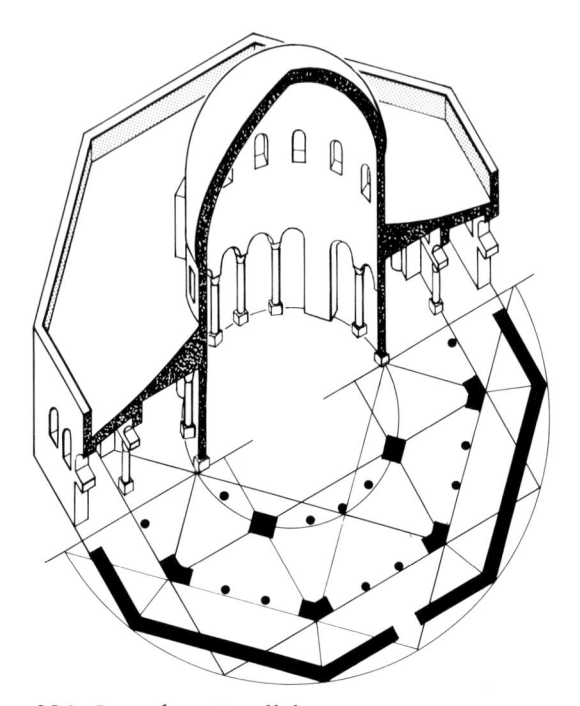

206 Jerusalem: Parallel perspective representation of the Dome of the Rock with ground plan and vertical projection (after K. A. C. Creswell).

138   *Design and Construction*

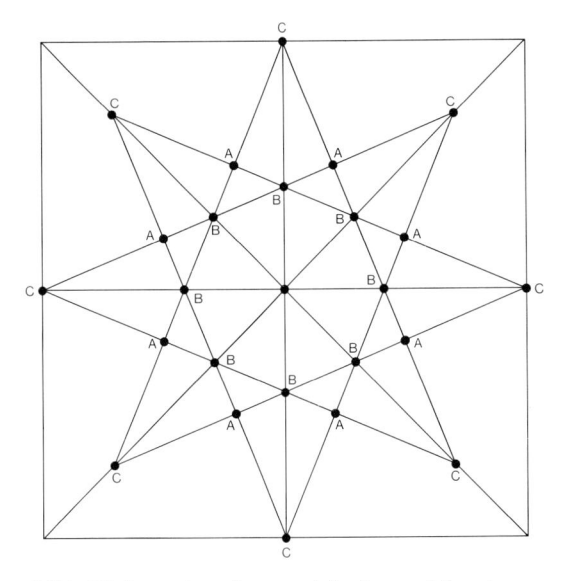

207 Eight-pointed star with the nodal points of the construction lines. When the eight-pointed star is used as the ground plan, these are the points of structural load.

of San Vitale in Ravenna (540 A.D.) and St. Simeon in Qal'at Sim'ān (end of the fifth century).[183]

On the basis of these various observations, it may be said that the eight-pointed star is fully developed in Byzantine architecture. If we compare the plans of San Vitale, St. Simeon, and the Dome of the Rock and the basic figure of the eight-pointed star common to them, we become aware of the multitude of variations that are possible within the framework of this figure and the ground-plan design based upon it, whereby one main principle is identifiable: the points at which the lines of the eight-pointed star cross are structural nodes (Fig. 207):

A The eight corners of the two basic squares
B The corners of the 'inner' octagon formed by these two squares
C The eight points of the star

Pillars, columns, and supporting sections of the wall are always found at A, B, or C.

In the preceding chapters we have attempted to fathom the oriental using the occurrence of the octagon as a parameter, and to follow the form and meaning of the octagon and the eight-pointed star in oriental architecture as far as East Asia. In so doing we have come across a variety of octagonal designs which have symbolic meaning that goes beyond their geometric-decorative shape.

We have called attention to the symbolic power of mathematical laws in relation to the enigmatic incommensurable, arithmetically indeterminable length of the diagonal of the square, which we come upon, along with the corresponding quantities $\sqrt{3}$, $\sqrt{4}$, $\sqrt{5}$, again and again in Islamic geometry and the decorative art it defines.

Eight is an interesting number for arithmetical reasons — this is the source of its special symbolic power in ancient times: the square of every uneven number above one is a multiple of eight plus a remainder of one! It was also found that all squares of uneven numbers above one differ from one another by a multiple of eight: $9^2 - 7^2 = 81 - 49 = 32 = 4 \times 8$.

Its symbolic meaning and its characteristics are of an esoteric nature. Beyond the seven spheres of the planets lay the eighth sphere, that of the fixed stars. In the Babylonian tower temples the deity abided on the eighth story — from here the eight developed into a number of

The Eight-pointed Star 139

paradise. In Islam there are seven hells but eight paradises; this term appears repeatedly in Persian literature. Eight angels carry the throne of God. Dante showed the 'church triumphant' in the eighth heaven, and in Buddhism one speaks of the eight-part path to nirvana. The pantheon of China recognizes the Eight Immortals, and the 8×8 figures of the I Ching confirm the significance of the number eight, as does Japanese tradition, according to which eight is the number that epitomizes infinity. This leads us back to that symbology of the hereafter which appears in the octagonal form of the mausoleums in the Indo-Islamic world.[184]

3 The Ground-plan Design: General Remarks

The question of how the design was developed and how it was translated into the practical building process arises for everyone who stands in front of one of the large medieval cathedrals. The question is all the more compelling when it pertains to such an unusual building as Castel del Monte, which belongs to the same period as the Church of St. Elizabeth in Marburg (begun in 1235) or the Naumburg cathedral (begun in the mid-thirteenth century) and yet differs so fundamentally from both of them in its nature and structure.[186] The attempt to form an idea of the design and building processes in the Middle Ages is complicated by the sparsity of written documents that have been handed down. We are fortunate, of course, to possess the sketchbook of an architect of that period, Villard de Honnecourt, but, as L. R. Shelby has rightly noted,[187] it is precisely the uniqueness — and thus the incomparability — of these sketches that makes them difficult to interpret.

Hans R. Hahnloser's excellent edition of the sketchbook[188] shows us that it consists mostly of instructions for the design of figured representations but also contains hints regarding building techniques. Some of the sketches were literally scratched onto the parchment with the

"Ars sine scientia nihil est."

Statement made by Jean Mignot at the conference of master builders of the cathedral of Milan 1391/92[185]

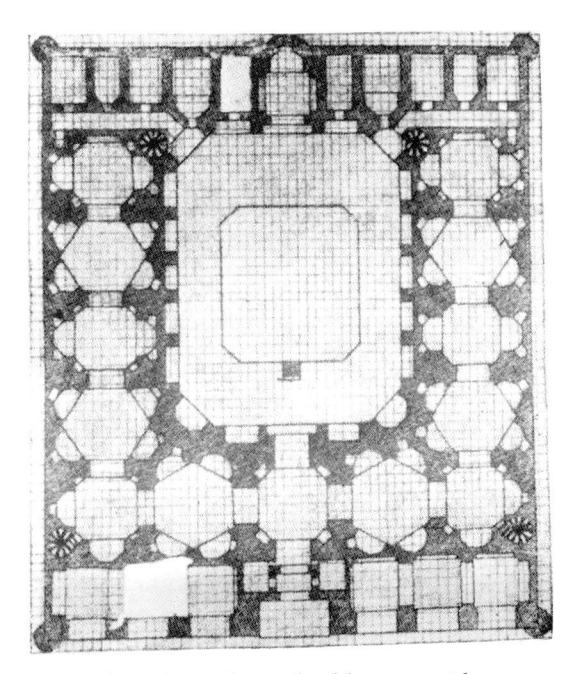

208 Plan of an Islamic building on grid paper (from Michell 1987, p. 114; see also G. Necipoğlu 1995, p. 8, Fig. 8), attributed to an Uzbek master builder from Buhārā; madrasa or caravansary. Tashkent, Uzbekistan, Institute of Oriental Studies, Uzbek Academy of Sciences.

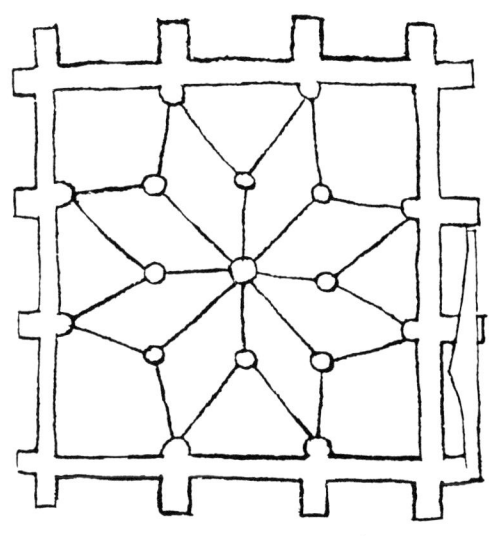

209 Sketch of an eight-pointed star vault from Villard de Honnecourt (from Hahnloser 1935, plate 41).

point of a compass and retraced or drawn with a pencil. Thus we know that people worked with the compass and the ruler. As far as the plans of church buildings are concerned, these may be regarded as merely design sketches, without being true to scale for use at the stonemasons' workplace. In the case of a figured drawing, a line grid was first made on the drawing paper.[189] Line grids like those we know from the Arab world (Fig. 208) were certainly also used for drawing the ground plans to scale. The dimensions from the drawing were then transferred to the workplace according to a given factor of enlargement. The dimensions can thus be expressed in terms of modular units (grids).[190]

Not found in Villard de Honnecourt's sketchbook are exact geometric constructions. His illustrations — in Hahnloser, top left on plate 30 or the eight-pointed star on plate 41 (here Fig. 209) — are not based on *geometric* constructions of either an octagon or an eight-pointed star. In this architectural geometry, exact geometry as determined by mathematics played a subordinate role. This is in agreement with other records: The earliest old-French treatise on mathematics and its application stems from the second half of the thirteenth century (1275/76).[191] It is difficult to imagine that such a geometrically complex building as Castel del Monte could have been designed on the basis of the knowledge that can be deduced from the sketchbook of Villard de Honnecourt. The diagonal of the square ($\sqrt{2}$), which played such an important role there, is found in Villard only in a supplementary table added by Master 2.[192]

Two hundred years after Villard de Honnecourt, two German master builders — Mathes Roriczer, who was the cathedral architect in Regensburg from about 1480 on, and the otherwise unknown Hanns Schmuttermayer — recorded their experience as artisans in two pamphlets (*Fialenbüchlein*, i.e., 'pinnacle booklet'). In 1487/88 Roriczer also published a 'German Geometry' (*Geometria deutsch*). L. R. Shelby[193] worked on these writings with the same care that Hahnloser devoted to the sketchbook of Villard, allowing us to form a lively picture of the two of them. Both begin their instructions with the construction of a large square in which, one after the other, seven further squares are enclosed, each rotated diagonally by 45° (Fig. 210). The squares are then placed next to one an-

The Ground-plan Design   141

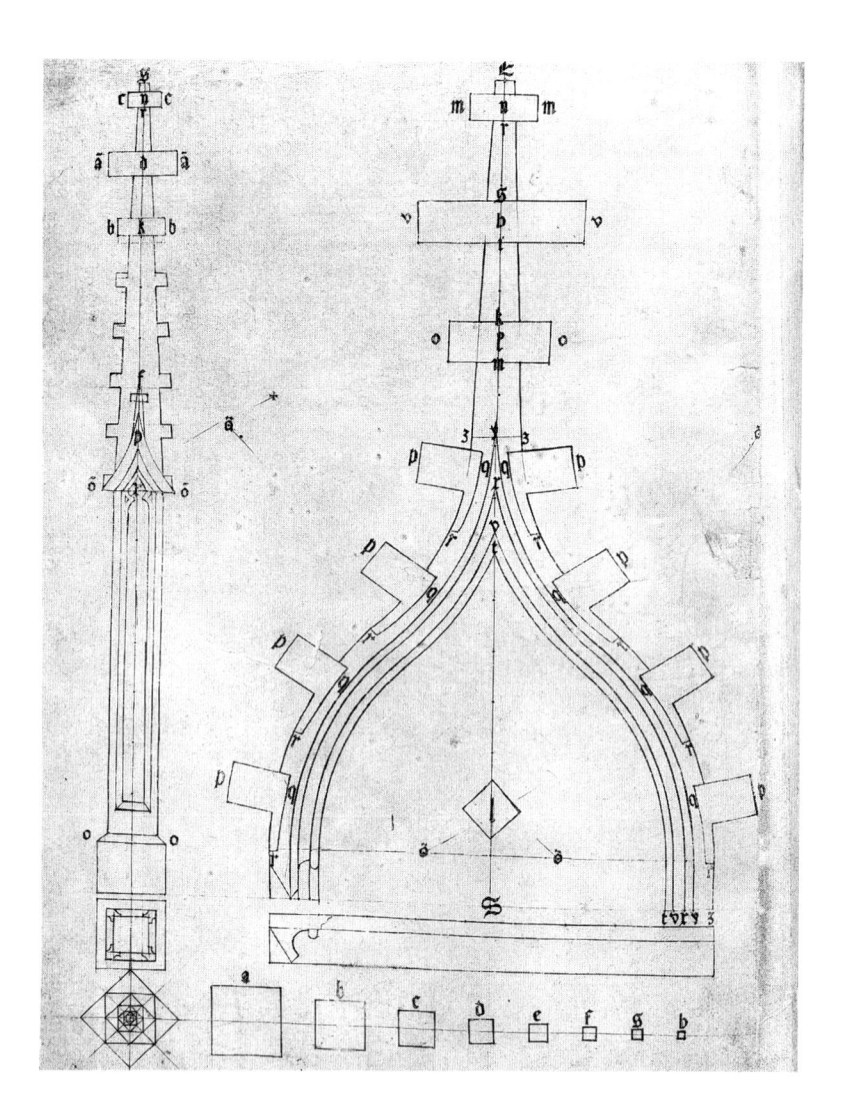

210 Drawing from Hanns Schmuttermayer's *Fialenbüchlein* (1484/86 and 1489). Nuremberg, Germanisches Nationalmuseum (cf. Shelby 1977, p. 144, plate II).

other and used as construction 'modules'. This is based on the well-known geometric method of bisecting the area of a square (see p. 116, Fig. 160). We also find the familiar motif of the 'rotating' squares (Roriczer, no. 1, folio 1, no. 2, folio 1^{v}),[194] which was no doubt an important point of departure for constructional considerations.[195]

In the *Geometria deutsch* Roriczer gives advice on a number of drawing techniques, for example, how to construct a right angle (Figs. 211 and 212) and from it a square, a pentagon, a heptagon, and an octagon. It is not a book *about* geometry, but rather a collection of some drawing tricks based on geometry, and it has little to do with mathematics per se. Thus, for instance, the description of the length of the circumference of a circle is extremely vague. The mathematically best approximation had already been given by Archimedes, whose treatise on measuring the circumference of a circle was translated

142   *Design and Construction*

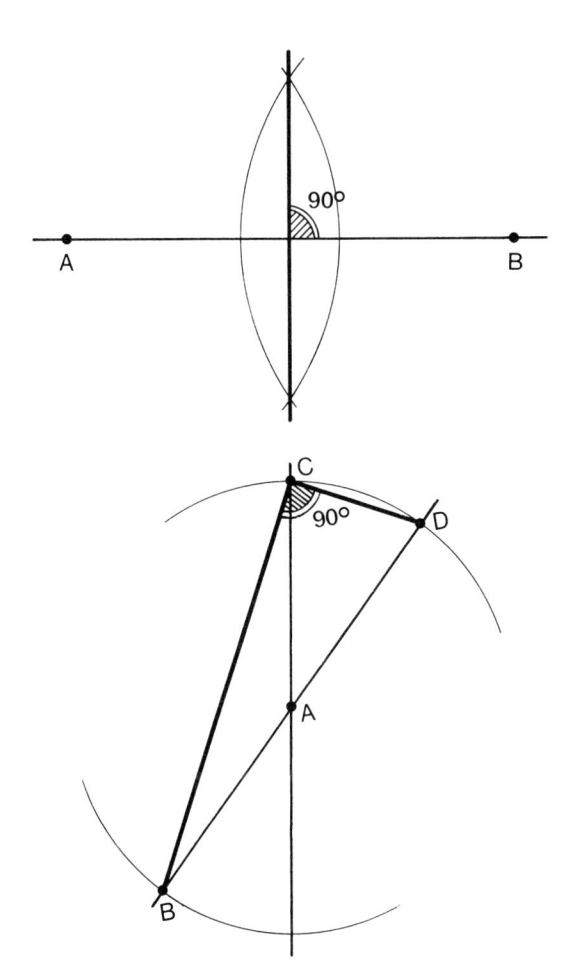

211 and 212 Two methods of constructing a right angle according to Mathes Roriczer (from Shelby 1977, pp. 85, 86).

from Arabic into Latin by Gerard of Cremona (ca. 1114–1187).

L. R. Shelby correctly called attention to the considerable difference between the completely pragmatic approach of Roriczer and Schmuttermayer to architecture and that taken by their contemporary Leon Battista Alberti in his ten books on the subject.[196] Like Augustine before him, Alberti compares architectonic proportions to musical harmonies.

Both the small pamphlets mentioned above and the sketchbook of Villard contain not so much fundamental theory as practical guidelines.[197] With the title 'Geometria *deutsch*' Roriczer surely intended as well to disassociate himself from scholarly Latin treatises on geometry, such as the already mentioned Latin translation of Archimedes by Gerard of Cremona. The great authority on the Gothic period, Otto von Simson, who wrote an impressive work devoted to Gothic architecture, writes: "The practical architectural geometry of the Gothic period made use, however, of extremely simple methods."[198] It goes without saying that the intention here is not to detract from the greatness and importance of this architecture.

The discussion that has been passed down to us between the builders of the cathedral of Milan and foreign advisors who had been consulted at the turn of the fifteenth century (among others, Heinrich Parler, Jean Mignot) gives an informative insight into the conceptions that medieval architects had regarding design and construction.[199] The discussion was related not so much to the plan, which had already been settled, as to the elevation and its structural problems.

We who wish to get closer to the form and thus the architectonic design principle of Castel del Monte must try to understand from what traditions the more complicated geometry that was applied there stems. Beginning in the Romantic period, art and architectural historians long believed that the Gothic stonemasons' lodges had kept a secret that had enabled them to create the admirable cathedrals, and that the designs for this architecture were based on secret rules of proportion that had been developed from the triangle or the square.[200] The existence of such a lodge secret was 'demythologized' by P. Frankl.[201]

Konrad Hecht has critically examined the supposed rules of proportion.[202] He correctly rejects the artificial

The Ground-plan Design 143

systems of proportions that have been imposed on the Gothic cathedrals again and again without any concrete, reliably substantiated leads. When Hecht comes to the conclusion, however, that the Gothic architects used a geometry of *dimensions* but not one of *proportions*, he is quite evidently going too far.[203] The "quadrature over the locus"[204] (Fig. 210) alone contains *proportional* geometry, because the smaller squares that are developed from the starting square follow one another in a geometrically *definable* proportion. The individual squares cannot be measured exactly because of their arithmetically incommensurable relationship to each other.[205]

The essence of *proportion* — as opposed to simple dimension *size* — is a *relationship* of dimensions, such as that represented by the golden section.[206] Proportion has an *aesthetic* quality in architecture — just as it does in music, which is related to architecture. In music it is expressed in the relationship of the tones to each other, i.e., in harmony. Everywhere in art, and thus also in architecture, it is not dimensions but rather dimensional *proportions* that determine the impression made on the one who is viewing or listening. This holds as well for Gothic architecture, whether it be the proportions of the pinnacles (*Fialen*) on the flying buttresses or the relationship of the height to the width of a nave or a portal facade — regardless of whether the proportions were arrived at geometrically or owe to the experience of master craftsmen alone and do not comply with any prescribed or later imposed system.

We owe the clearest and most convincing delineation of the design and building process in the Gothic architecture of central Europe to François Bucher.[207] He develops the steps involved in the design and realization of the construction plans, from rough sketches, to preliminary plans, to detailed plans, and so on, up to the master drawings and their transfer to the 'drawing floor', including all practically required data. Although the material on which Bucher's description is based stems mainly from the period between 1350 and 1572, it is surely expressive of the situation during the preceding century.[208]

My attempt to reproduce the ground plan of Castel del Monte starts with the geometric relationships that are *visible* — the eightfold rotation symmetry, the eightfold axial symmetry, and the threefold homeomorphic sym-

213 Castel del Monte, Apulia:
Ground plan with eight-pointed star drawn in.

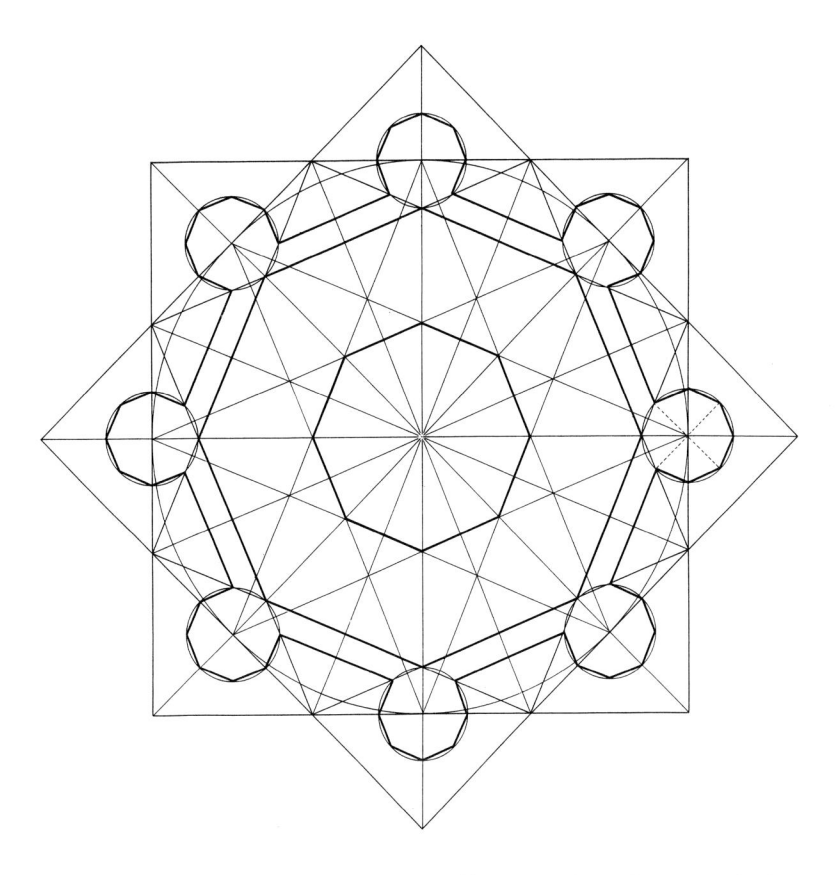

metry between the geometrically 'similar' forms of the inner court, the outer wall, and the eight corner towers. Alone the exact right angle at which the sides of the towers meet the walls of the large main octagon (see p. 163, Fig. 233) assumes a systematic geometric plan that was reconstructible on the basis of the indicated geometric conditions. Thus, no attempt of the sort criticized by K. Hecht was made to impose on the ground plan an abstract 'ideal plan' that bears no concrete relation to the building itself — something that has erroneously been assumed on occasion. The plan illustrated here in Fig. 213 goes far beyond the geometric knowledge and experience that we can presume, based on the sketchbook of Villard de Honnecourt and even on later records from the fifteenth century.

Nevertheless, there is contemporary evidence from quite another field that confirms the correctness of the ground-plan structure deduced for Castel del Monte. It concerns one of the so-called Portolan maps, the *Carta Pisana*. This nautical chart of the Mediterranean dates back to the third quarter of the thirteenth century and can be found in the Bibliothèque Nationale de Paris.[209] A detail of the chart is reproduced here as Fig. 214, showing

The Ground-plan Design   145

214 Detail from the *Carta Pisana*, a so-called Portolan map from the third quarter of the thirteenth century (Paris, Bibliothèque Nationale, Cartes et Plans, Reg. Ge B 1118).

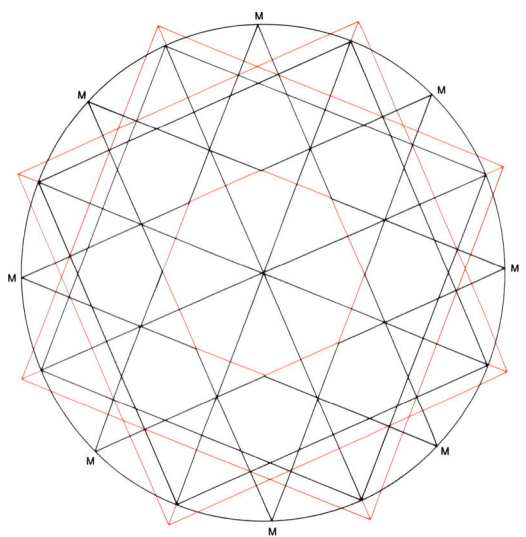

215 Schematic drawing of the compass rose in the *Carta Pisana*. The two outer squares plotted in red correspond to the ground plan drawing of Castel del Monte in Fig. 213.

the two intersecting squares for the construction of a compass rose. The result is a geometric figure that is identical to our ground plan of Castel del Monte (Figs. 213 and 215). The compass rose is congruent with the plan except for one small variation: In order that the intermediate wind directions do not extend further than the main directions, the points of the former are somewhat shortened and end on the same circle as those of the *main* directions (Fig. 215).

In his excellent book about the planning of newly founded Florentine cities, David Friedman makes the following comment with regard to the compass rose on the *Carta Pisana*: "The figure of the compass rose is an obvious representation of proportions characteristic for the structure of the new-town plans."[210] Friedman is referring here to the proportional arrangements, based on a table of chordal values from Leonardo of Pisa's *Practica Geometriae*, that were applied in planning the new Florentine towns. In addition, he draws a connection between the depiction of the compass rose and the astrolabe, which represents the celestial bodies.[211]

Here the repeatedly substantiated symbolism of the number eight is combined with the basic concept of the

146   *Design and Construction*

four, or eight, directions in the cosmological dimension of compass rose and astrolabe. Moreover, the role of Leonardo Fibonacci as one who imparted the knowledge of Arab trigonometry is clearly recognizable.

4 The Geometric Structure of the Plan

"One can hardly overestimate the depth of geometric imagination and inventiveness that these patterns reveal. Ornamental art is without a doubt the oldest known manifestation of higher mathematics."

Hermann Weyl [212]

Figure 213 illustrates the geometric structure of the plan of Castel del Monte. The essential characteristic is the tight connection of the inner court as the center of the construction with the middle points of the eight towers. These lie at the points where the extensions of the court sides intersect, originally the sides of two 'rotating' squares.[213]

With regard to the development of the octagon and the eight-pointed star, the compass rose of the Portolan map — the *Carta Pisana* (Figs. 214 and 215) — is designed in the same way as the plan of Castel del Monte. The intersections of the extended sides of the squares mark the points of the eight main wind directions. Between these lie the corners of the two large 'rotating squares', which were made somewhat smaller here than in the regular construction of the plan (Fig. 216) in order that the corner points meet the circumscribing circle on which all points of the star lie.

This relationship between the middle point of the two 'rotating' squares — i.e., the center of the citadel courtyard — and the corners of the star — i.e., the middle points of the citadel towers — is the main feature of this symmetrical group. We will find the same relationships again in a sophisticatedly elaborated example of Islamic ornamentation from the Alhambra in Granada.

If we first take another look at the plan drawing of Castel del Monte (Fig. 216) and allocate to each of the two large squares corner squares, whose diagonals are bound by the points of intersection of the two large squares (Fig. 217), the resulting small squares are also characterized by the fact that their middle points represent the centers of the corner towers.

216 Castel del Monte, Apulia: Construction design of the ground plan.

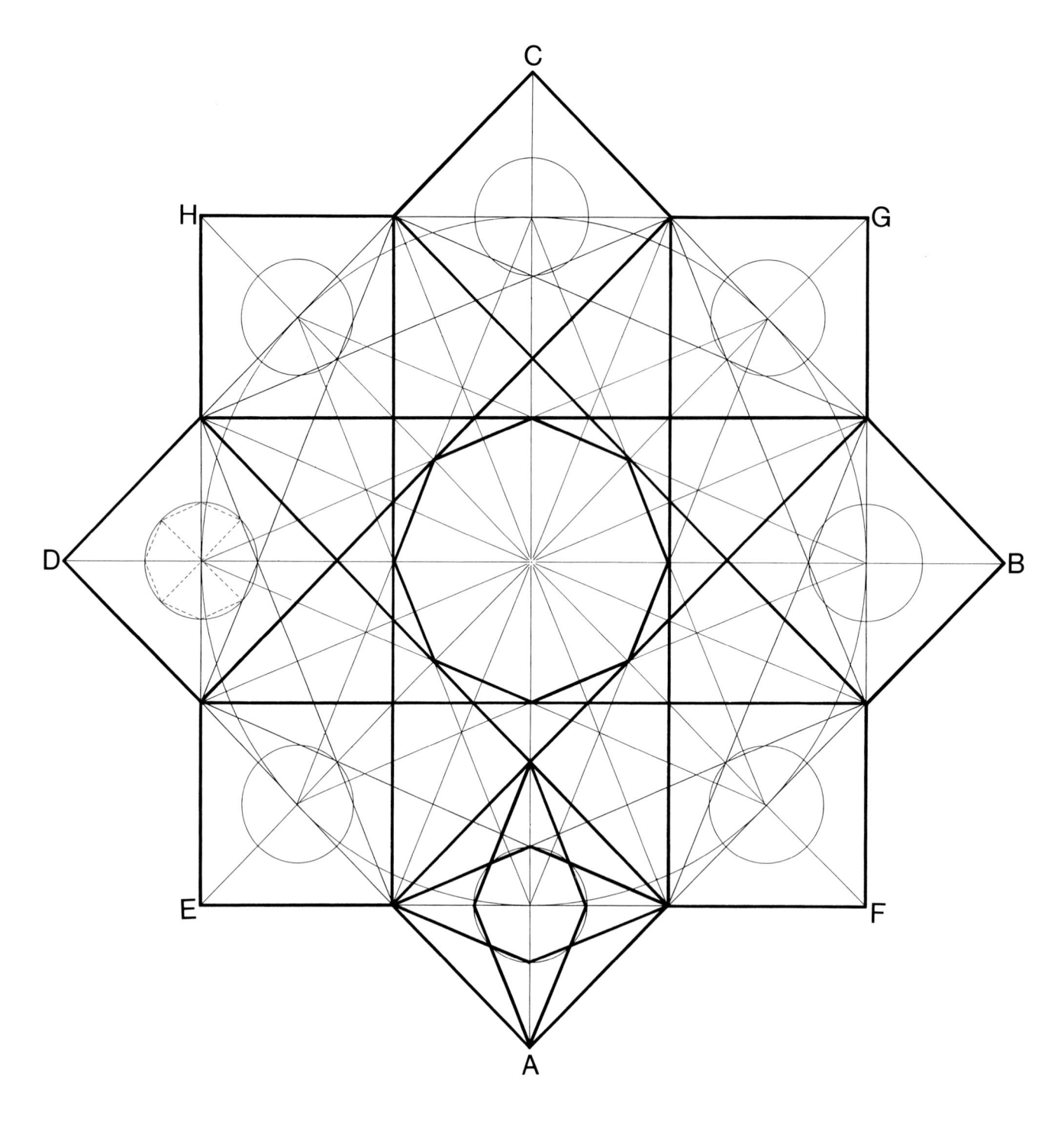

Without being aware of my construction (Fig. 217), Rafael Pérez Gómez, who has occupied himself for quite some time with the symmetrical groups of decorations in the Alhambra (thirteenth to fourteenth century), made the drawing reproduced here in Fig. 219, based on the ornament in the Alhambra, Granada, El Palacio de Comares, El Salón del Trono. The surprising identity is no coincidence. Rather, the drawings that were done indepen-

217 Plan of Castel del Monte. The centers of the towers are in the middle of squares that are defined by the two large intersecting squares.

148   *Design and Construction*

218 Mosaic from the Villa of Piazza Armerina in Sicily (from Carandini et al. 1982, p. 246, Fig. 146, and front cover of volume of plates). The late Roman mosaic here shows the same arrangement of the side squares as seen in Figs. 217, 219, and 220; in Castel del Monte their midpoints determine the centers of the towers.

dently of one another (Figs. 217 and 219) confirm — in connection with the exactly identical basic geometric motif of the compass rose in the *Carta Pisana* — the frequency of this geometric construction in the Islamic world. Its use as the basis for a decoration, on the one hand, and as the outline drawing for a building, on the other, is not surprising, as the underlying geometry serves the two closely related arts in the same way.

The Geometric Structure of the Plan 149

Characteristic of the two drawings is the close and clearly marked relationship between the center of the inner octagon, the center of the building as a whole, and the middle points of the eight corner towers. In the decoration that R. Pérez Gómez used as a basis, these tower centers are marked by smaller octagonal ornament centers corresponding to the eight corner towers,[214] but rotated by 22.5° (cf. Fig. 220).

219 This drawing by Rafael Pérez Gómez shows the detail of an ornament from the Alhambra that is reproduced in Fig. 220. The ornament is framed by a basic structure determined by the artist, which is identical to Fig. 217. The centers of the 'satellite stars' surrounding the main star correspond to the centers of the eight towers of Castel del Monte.

150   *Design and Construction*

220 Granada, Spain:
Alhambra, El Palacio de Comares, El Salón del Trono (thirteenth/fourteenth century). The sketch superimposed on the ornament brings out the two intersecting squares as well as the corner squares, whose touching points mark the centers of the 'satellite stars'. The centers of the squares lie where the centers of the corner towers are located in the plan of Castel del Monte (cf. Figs. 217 and 219).

The same ornament from the Palacio de Comares, El Salón del Trono, reproduced here photographically in Fig. 220, permits just as convincingly another construction of the 'inner court' and the tower centers in the middle points of the corner squares surrounding the two 'large' rotating squares. Here a certain constructional variation can be seen: The eight 'outer' centers (*sinos*) are located not at the middle points of the octagonal corner towers

The Geometric Structure of the Plan 151

but at the meeting points of two corners of the eight surrounding squares, which are shaped as virtual eight-pointed star 'satellites'. To conclude, we may say that both of the decorative constructions and Castel del Monte itself are — mathematically expressed — actually D 16-type *symmetrical groups on the plane*.[215]

The form of these symmetrical groups designed on the plane conforms entirely to the already observed *two-dimensional* concept of *spatially* appearing structures of Islamic buildings — for instance, the belts of the cupolas in Córdoba (Figs. 178 and 181); they might be termed *pseudospatial* structures. For this reason, however, there is no relationship that can be directly inferred from the plan between it and the building's height; the building represents a monohedron, the height of which is independent of the ground plan.[216]

It is surprising to see that the geometric nodal points, representing the centers of the ornament groups shown from the Alhambra, called *sinos*, coincide with the 'power points' of the eight-pointed star system shown in Fig. 207. Such a coincidence is mandatory for the plan of Castel del Monte, just as it is for the eight-pointed star cupolas at Córdoba.

If more proof were necessary of the fact that the design plan of Castel del Monte is based on a geometric construction, this would be it. Moreover, it is not a fortuitous ad hoc construction, but a geometric figure from regions where — as evidenced by the Portolan map and the Alhambra — "advances in surveying which had come to Europe with the introduction of Arabic learning and mathematical instruments" were frequently used.[217] In addition, this figure comprises a wealth of symmetrical relationships — in particular 'similarities' — which the Russian architectural historian M. S. Bulatov, in his book *Geometric Harmonization in the Architecture of Central Asia in the Ninth to Fifteenth Centuries*, regards as especially characteristic for Islamic architecture.[218] Bulatov examined the geometric proportions of a large number of central Asian buildings dating from the eleventh to the fifteenth century. He came to the conclusion that, in their endeavors towards the harmonious design of buildings in the region most important for Islamic architecture, the architects took the following elementary geometric figures as their basis:

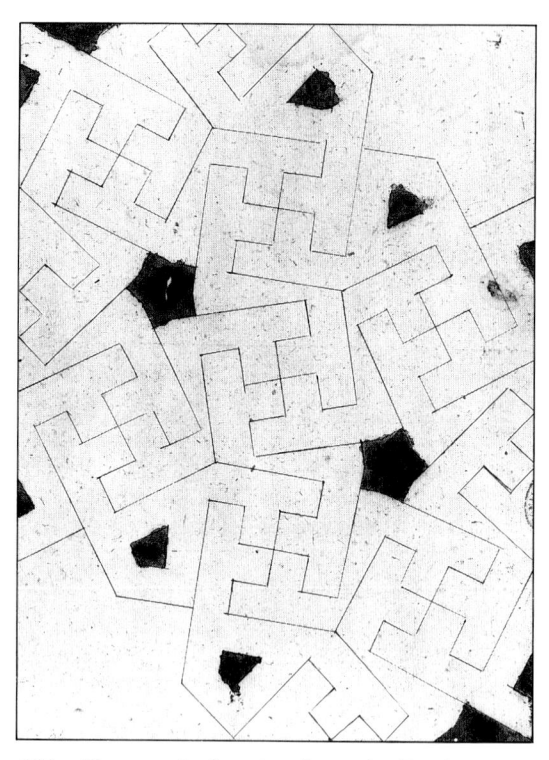

221 Geometric drawing from the Persian manuscript by Abū l-Wafā' al-Būzaǧānī (Paris, Bibliothèque Nationale, Persian manuscript no. 169). Gülru Necipoğlu refers to this sketch in her book, Fig. 113a (p. 149), as a "repeat unit for a star-and-polygon pattern with swastikas." It is a complicated geometric figure that can be repeated as desired on all sides.

222 Grid diagram on which Fig. 221 is based (drawing by Elizabeth Dean Hermann, in Gülru Necipoğlu 1995, p. 149, Fig. 113b). A very ingenious geometric construction is perceivable, uniting trefoil, square, and pentagon in a most artful symmetrical ornament.

- The square and forms derived from it
- The equilateral triangle and forms derived from it
- The half square (rectangle)
- The division of a distance by means of a 'geometric progression' ('golden section'; cf. also the 'Fibonacci sequence')

Bulatov's conclusion rested not only on Euclid's geometric theses but also on expositions by great Arab scholars, such as Abū Naṣr al-Fārābī (ninth century A.D.), Muḥammad al-Ḫwārizmī (ninth century A.D.), Abū 'Ali al-Husain ibn 'Abdallāh ibn Sīnā (Avicenna; 980–1037), and Ġiyāṯ ad-Dīn al-Kāšī (fifteenth century A.D.; author of *The Key to Mathematics*).

Above and beyond this, it is to Bulatov's special credit that he based his interpretation of architectural geometry on a treatise dating from the eleventh century written in Persian by an anonymous author: 'On That Which Artisans Must Know About Geometric Constructions'. It is apparently a transcription of lessons by Abū l-Wafā' al-Būzağānī (940–998 A.D.).[219]

Treatises of this type continued to appear later as well. For example, the important astronomer Ġiyāṯ ad-Dīn al-Kāšī from the observatory of Uluġ Beg in Samarqand wrote the above-mentioned work entitled *The Key to Mathematics*, a chapter of which is devoted to problems of surveying in architecture.

The reproduction of a drawing after al-Būzağānī (Figs. 221 and 222) reveals how far this 'practical geometry' of the tenth century was ahead of the geometry of the central European lodge books. This will not surprise anyone who is familiar with the history of Arab mathematics. Al-Būzağānī was one of the last translators and commentators of Greek mathematicians such as Euclid and Diophantos. He wrote an abridged version of Ptolemy's *Almagest* and composed a textbook of applied geometry. He made a decisive contribution to the development of trigonometry and calculated a sine table of 30' to 30' exactly to eight decimal places.

There is no doubt that at that time and in those geographic areas there was a considerable difference between the knowledge a learned mathematician possessed and that of the artisans and workers in the building trade, who — using it to decorate their mosques, mausoleums, and palaces — transformed the high art of geometry into

The Geometric Structure of the Plan 153

visible testimony to the infinity of the cosmos. It is the Islamic way of venerating the divine that found its highest form of expression in this geometric abstraction.

From the geometric drawings in the treatise of al-Būzaǧānī we can deduce the high level of *trade* geometry that was realized in the Islamic architectural works and their ornamentation. Present-day mathematicians only recently established that in the creation dating from the late period of Islamic art, the Naṣrid Alhambra in Granada, all 17 mathematically possible symmetrical groups on the plane are represented. The mathematical proof that only 17 are possible was produced less than 100 years ago![220]

Bulatov's interpretations of the construction of Islamic buildings show us what an important role was played by the choice of the *correct proportions* for the architectural dimensions and the *similarities* of the geometric forms. We are reminded of the similarities (homeomorphic symmetries) described in the construction drawing of Castel del Monte, but also of even earlier examples — for instance, in Castel Ursino (p. 45, Fig. 48). Conspicuous as well is how often the proportional relationships based on the incommensurable quantities $\sqrt{2}$, $\sqrt{3}$, and $\sqrt{5}$ are used. A representational example of the relationship between the side of the square and its diagonal ($\sqrt{2}$) is shown here in Fig. 223; it is supplemented by a further example from Aḍerbaiǧān with the application of a Pythagorean triple and the $\sqrt{2}$ (Fig. 224).

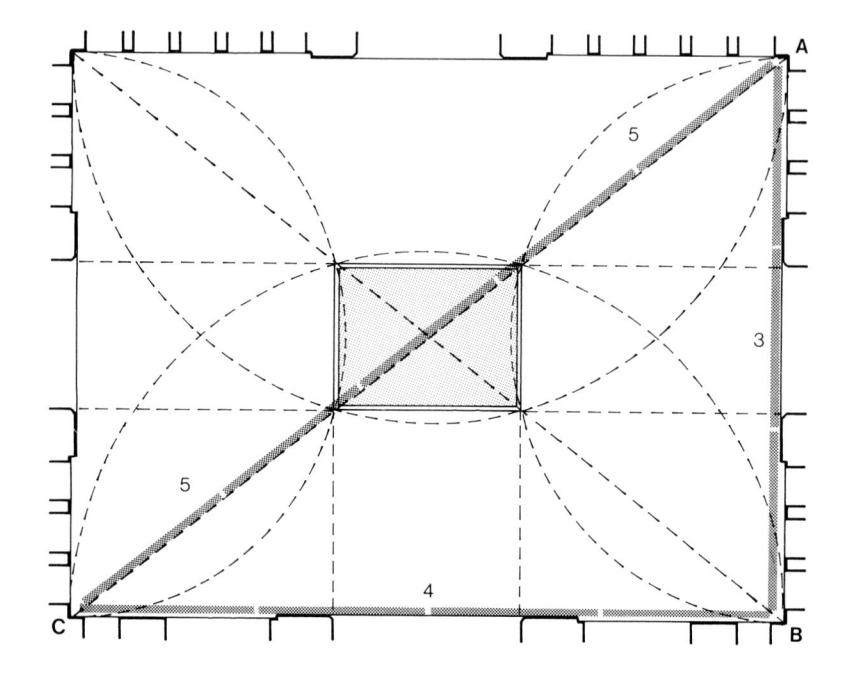

223 Iṣfahān:
Ground plan and layout of the central court of the Šāh mosque. The points of intersection of the semicircles above each court facade coincide with the diagonals and define the corners of the basin in the center. The sides of the triangle ABC are in the ratio of 3 : 4 : 5 and thus form a Pythagorean triangle. The width of the central basin is the same as that of the east, west, and north iwan; its length corresponds to the width of the south iwan.

154   *Design and Construction*

224 Mausoleum of Šaiḫ Dursun in Aderbaiğān (1403–1404). Between the total width of building A and the inside diameter *a* there is the correlation A = √2 *a*. Moreover, the Pythagorean theorem is manifest in the cross-section of the octagonal pyramid: half of the base is in a ratio of 3 : 4 to the height, resulting in a quantity of 5 for the line of intersection of the pyramid side: $3^2 + 4^2 = 5^2$ (from Bulatov 1978).

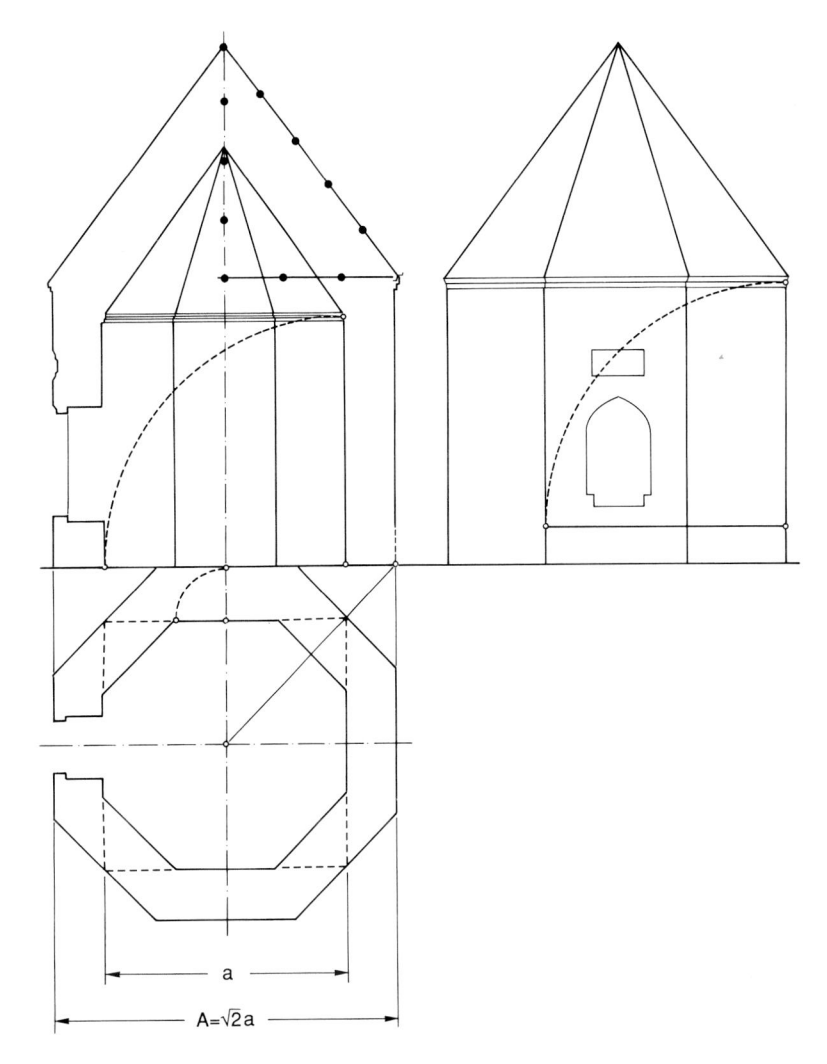

Our reflections and comparisons have shown that the geometrically complex design plan of Castel del Monte is indebted to Islamic tradition, regardless of how differently the construction in detail and the plastic design of the building may appear. In the cultural melting pot that was the kingdom of the Two Sicilies this is not surprising; rather, it adds another facet to the overall picture. We witness the felicitous amalgamation of plastic Greek-European architectural tradition with geometric design structures from a cultural complex that Frederick II knew intimately and admired. Castel del Monte is thus an incomparable example of the mathematical idea of aesthetics. Islamic geometry carried on the legacy of geometric thought from the Greeks.

Thus we observe repeatedly that all attempts to explain the shape of Hohenstaufen four-wing constructions as well as the complicated octagonal plan of Castel del Monte based on central European traditions alone

The Geometric Structure of the Plan 155

prove to be untenable. There *are no* forerunners of these strictly geometric and completely regular symmetrical plans in central Europe. Romanesque and early Gothic octagonal designs have a totally different structure and they lack the geometric form of the eight-pointed star characteristic for Castel del Monte and the examples shown from the Islamic area, with its fixed connections between the main center and the secondary centers of the star points. In addition, Islamic influence is nothing uncommon in the region of southern Italy and Sicily. However, it is frequently and all too easily overlooked because of the virtually *canonical* viewpoint of the observer in central Europe. From the position of Sicily, which was under Islamic rule for two and a half centuries and still maintained very close ties with northern Africa during the Norman period, the picture is different.[221] As early as 1927, Ernst Kühnel called attention to Islamic influences in the Romanesque art of Italy. An example of how forms were taken over directly, for instance, is the cathedral of Troia in Apulia (p. 66, Fig. 91), which lies in the countryside that Frederick II loved.

The realization of the building design, which was new and geometrically complicated for the Cistercian lodges, most certainly required special skill and practical experience on the part of the master builders and craftsmen. The octagonal arrangement of the trapezoid rooms was unusual compared with the series of square, ribbed groined vaults in a straight sequence of yokes,[222] like the particularly impressive one in Castel Maniace, built at almost the same time (Fig. 225).

The Cistercian ribbed groined vaults on the two stories of Castel del Monte appear not in a continuous sequence but singly, in each of the trapezoid-shaped rooms. On the sides of each individual yoke the beginnings of the ribs for a neighboring yoke can be seen, but — as there is no neighboring yoke — these come to a dead end in the adjacent masonry (p. 96, Fig. 136). Compared with each other, the dimensions of the yokes are amazingly identical, although they differ from one story to another, owing to the thicker walls on the lower floor. The yoke squares on the upper floor are thus distinctly larger (7 m as opposed to 6.40 m), because the inside distance between walls is greater there. This is evidenced by the smaller distance between the columns bounding

225 Syracuse, Sicily: Castel Maniace. View of one of the restored yoke sequences.

156 *Design and Construction*

the yoke and the inside corner of the trapezoid next to them on the upper floor (pp. 95/96, cf. Fig. 135 with Fig. 136). It is clear from this that not the yokes, but the wall lines as determined by the ground plan were the primary construction element of the building.

On the main octagon, the central part of the citadel, the following are geometrically defined:

1. The inner alignments of the outer walls on the lower floor.
2. The inner alignments of the outer walls on the upper floor. They are recessed in accordance with the lesser thickness of the outer walls compared with those on the lower floor but follow the alignments of the (not visible) insides of the octagonal towers (see Fig. 245).
3. The outer alignments of the outer walls on both floors.
4. The outer alignments of the inner (court) walls on both floors.

Not defined by the geometric plan are the inner alignments of the inner (court) walls on both floors. The regular wall of the large octagon, determined by the 'rotating squares', is the belt that holds everything together. Its amazingly precise measurements confirm the central function of the large octagon.

The complicated plan itself called for an exact final plan which served as the basis for the transfer to the building site. Trigonometric methods and instruments were available for this transfer. The dimensions of the various building components were closely connected with and dependent upon one another, owing to their geometric relationships. Nevertheless, the 'dimensional accuracy' of Greek buildings of the classical period should not be expected in a medieval building of squared stones; with their most precise art of handling marble the Greeks were able to produce exact dimensions — not only on level surfaces but also with complicated contours. We will have to reckon with deviations. In his analyses M. S. Bulatov found deviations on the order of 2%, in some cases even of 5–8%.[223] A. Kottmann mentions deviations of 1–3% but points out correctly that such percentages have only limited value, since with short distances the permissible errors are too small, while with long distances they are much too large.[224] To the best of our knowledge, the deviations that must be taken into account for Castel del Monte are relatively minor.

The Geometric Structure of the Plan 157

5 *The Geometric Construction and Measurements*

In our consideration of the cupola vaults of Córdoba we noted that the eight points of the star had a structural function: the weight of the cupola was concentrated in them — at least optically. They were the lines of force that transferred the weight onto the columns or pillars in the inside corners of the octagon below — similar to Gothic transverse and diagonal ribs. In San Vitale at Ravenna as well, the points end at structural nodes: the eight supporting pillars of the outer wall. The same thing can be seen in the plan of the Dome of the Rock in Jerusalem (p. 137, Fig. 205). The eight star points meet the eight pillars that carry the massive roof; at the same time, they are closely tied to four inner nodal points of the star, i.e., with the corners of one of the underlying main squares. The functional character of the eight-pointed star motif that is expressed here continues, as we have seen, down to the variations of the basic motif of the wall decoration in the Alhambra (p. 148 ff., Figs. 219 and 220), which exhibit a clear geometric relationship between the midpoint of the ornament group and the midpoints of the secondary centers. The architect of Castel del Monte was fully aware of this functional character of the eight-pointed star motif and disregarded *everything* that did not appertain to this geometry of the octagon and its symmetrical group: The midpoint of the ornament group corresponds to the center of the overall construction; the midpoints of the secondary centers correspond to the midpoints of the towers (Fig. 226).

It may be assumed that the emperor himself, who participated so actively in discussions of mathematics,[225] was also interested in its application to architecture, which he valued just as highly. There is evidence, for instance, of his contribution to the designs for the bridge citadel at Capua.[226]

In Castel del Monte the geometry of the octagon and the eight-pointed star is represented purely and consistently; there are no additions or annexes that might have obscured the geometric structure. This is what constitutes the aesthetic effect that is felt by the observer even without knowledge of the 'why'. There is no wall, inside

226 Planimetric aerial photograph of Castel del Monte with drawn-in eight-pointed star.

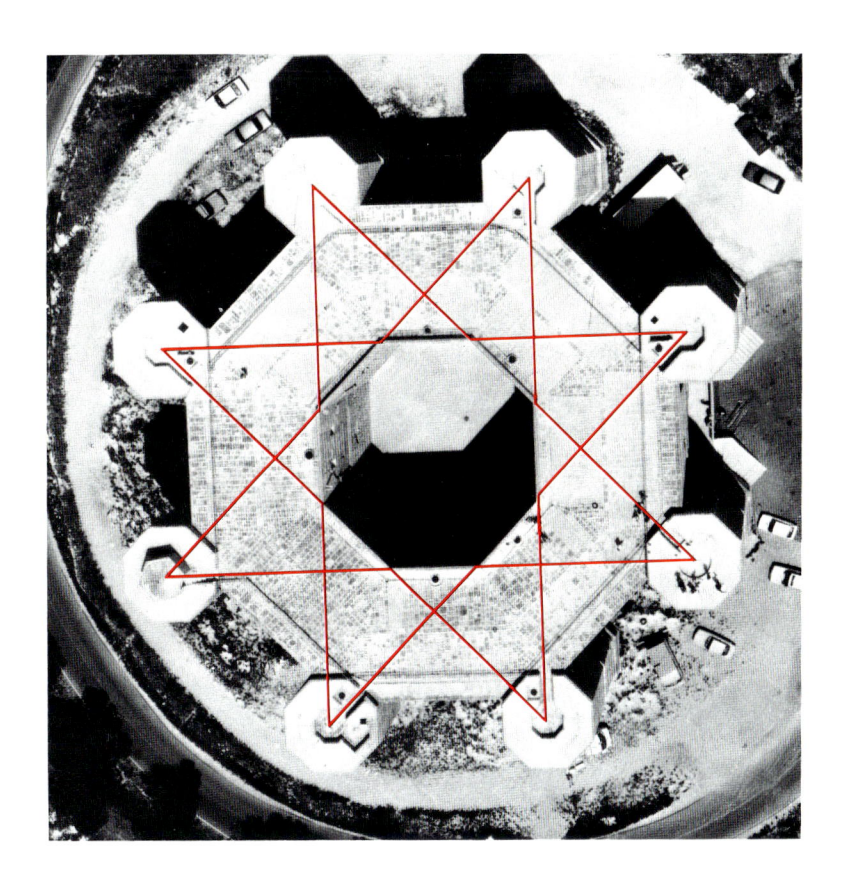

or out, that does not conform with pure geometry. Alone the splendid portal resembling a triumphal arch deviates from the strict geometry, thus accentuating it and making it all the more effective.

The geometric-stereometric structure is a reminder of what we have already observed in connection with other of Frederick's buildings: their 'plastic' substantiality, revealed in the repression of two-dimensional ornamentation and in the pleasure taken in large, clear surfaces as the delimitation of plastically perceived building elements. We noted that even the interruption of the surfaces through windows or portals was accomplished with the plastic means of depth gradation. All of this reached a zenith at Castel del Monte, in which the architectonic concept was brought into accordance with the mathematical-geometric theory. In what follows we will attempt to reconstruct graphically the design process.

On the 29th of January 1240, in Gubbio, the emperor commanded a man whom he personally trusted, R. de Montefusculo, to have a composition floor of lime mortar, stone, and other suitable material prepared for the citadel that he intended to erect near Santa Maria del Monte.[227]

With such a geometrically complicated ground plan, which in addition also has very precise proportions, a design sketch, possibly on grid paper, was indispensable. It is hardly possible to reconstruct the transfer of the final drawing onto the building site without carrying out excavations at the site itself. Even then, it would surely be difficult to discover measuring points since, if they did exist, they would have been obliterated by the subsequently erected building.

From time immemorial, the delimitation of the base line for a building had been the first important ceremonial act. For temples and official structures in particular, the orientation of the center line in a given direction was itself part of the construction drawing. Irrespective of the meaning of the word 'actractus',[228] the fact remains that transfer of such a complex ground plan from the drawing to the building foundation was possible only with the use of exact geometric techniques. Examples were the astrolabe and the surveyor's quadrant (see note 198) and the stretching of ropes. The highly respected rope stretchers, in Greek *harpedonapten,* existed as far back as in ancient Egypt.[229]

It is assumed that the medieval builders were also familiar with complex geometric methods and were able, for example, with the help of a rope divided by knots into twelve equal sections (Fig. 227) — according to the features of a triangle later named after Pythagoras with sides 3, 4, and 5 — to construct a right angle only in the plane. If one defines one point as the beginning and end of the rope and bends it at the fourth and ninth knots, the result is necessarily a right angle between the shorter sides 3 and 4, the legs of the triangle, which obeys the rule $3^2 + 4^2 = 5^2$ ($a^2 + b^2 = c^2$). The rule was known to the Babylonians as far back as the first half of the second century B.C.[230] It cannot be proven without a doubt that the Egyptians and the Greeks used this method to construct the right angle, but based on various sources it is highly probable that they did.[231]

Von Simson postulates that this method of construction was also used to measure the plans for Gothic cathedrals.[232] Geometric techniques such as those recommended by Roriczer (cf. p. 143, Figs. 211 and 212) for constructing a right angle were used to draw the design on parchment.[233] It is possible that the design was plotted on

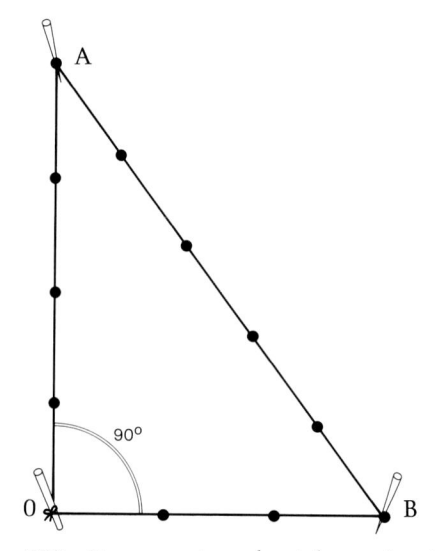

227 Construction of a right angle with the aid of a rope divided by knots into twelve equal sections. This construction is based on the Pythagorean theorem, the principle of which was already known to the Babylonians in the first half of the second century B.C. The ends of the rope are firmly fastened with a peg at point 0. A moveable peg is inserted at the fourth knot and stretched from point 0. Next the direction for the section 0A is determined, and point A is established. Then the second moveable peg is inserted at knot 9 and stretched from A and 0. The result is a right angle at 0, between 0A and 0B.

a grid like the ones we know from Villard de Honnecourt, but also from Islamic drawings (p. 141, Fig. 208). M. S. Bulatov, too, presupposes the use of such grids.[234]

After the right-angled triangle X 0 Y has been plotted, it is easy to complete a rectangle with sides of 3 and 4 units in length. If to the long side of this rectangle we add another, smaller rectangle, with sides of 1 and 4 units in length, we have a square with sides 4 units long (Fig. 228).

Step B (1) (Fig. 229): From point 0, the starting point of our construction, we describe a circle with a radius of 4 units and quadruple the original square, thus obtaining a large square I II III IV which contains the circle just described.

Step C (1) (Fig. 229): If we draw the diameters and diagonals of the square through the center of the circle and the square we obtain eight intersecting points 1 to 8. We have thus created unawares the same figure as the Moroccan craftsman did to construct his eight-pointed star (p. 132, Fig. 191).

Step D (1) (Fig. 230): If we connect each of the four tangent points of circle and square with each of the two opposite points of intersection of the circle with the diagonals of the square — as did the Moroccan construction draftsmen — we obtain the eight-pointed star in its geometrically defined form and with the two 'rotating' squares, which we have repeatedly encountered as a principal figure.[235]

Step E (1) (Fig. 231): In the interior of the figure a smaller octagon can be seen, which corresponds to the inner court of Castel del Monte. If we extend the two crossed squares inside it in all eight directions by a rectangle each whose sides are square side times half square side, or, in other words, if we enlarge each of the two squares inall eight directions by a rectangle having half the area of the square, we then have the boundary line of a larger octagon, which corresponds to the inner alignment of the outer wall of the large citadel octagon. The distance of the intersecting points of these eight inner alignments of the outer wall from the points of the star forms the radius of the circle that circumscribes the eight corner towers.

Step F (1) (Fig. 232): Now we construct the radius of the circles that circumscribe and thereby delimit the octagons of the eight outer towers.

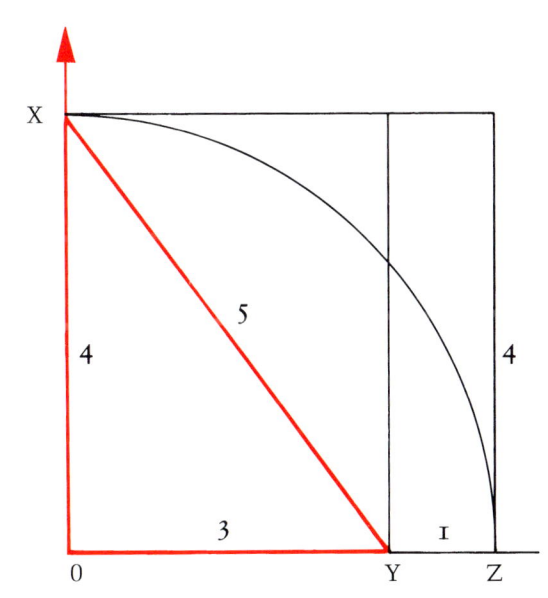

228 Step A: According to the principle elucidated in Fig. 227, a right angle Y 0 X is defined and the section 0 X is laid in the desired direction. The right triangle Y 0 X is expanded to a rectangle and this in turn, with the addition of a narrow rectangle 1 × 4, to a square. A circular arc with the radius 0 X is described around 0.

The Geometric Construction and Measurements 161

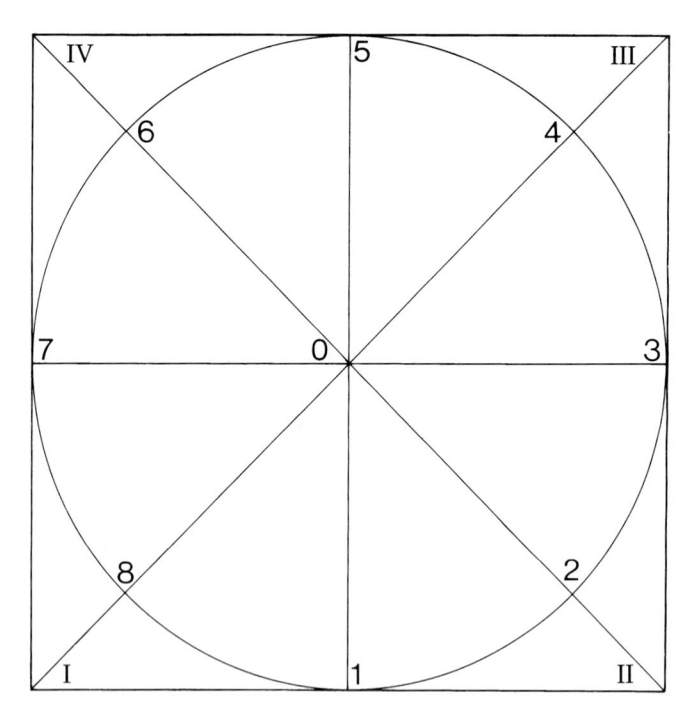

229 Steps B/C: The original square with the quarter arc is quadrupled to a large square I II III IV with an inscribed circle that cuts the diameters and the diagonals of the square into 1 through 8.

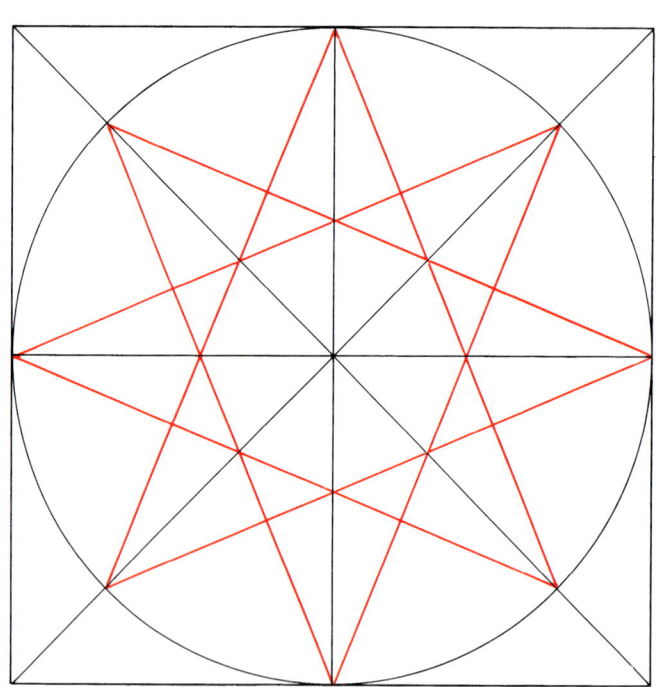

230 Step D: Each point at which the circle meets the square, in 2, 4, 6, and 8 (step C), is connected with the two opposing points where the circle intersects with the diagonals of the square (red lines). The result is an eight-pointed star that defines the interior courtyard.

Step G (1) : Fig. 233 shows the construction of the octagonal towers in detail. The outer alignments of the large octagonal wall start from the two octagon edges of the towers which face the middle of the building (K, K'). At these points they form right angles with the two adjacent sides of the tower octagons.

Until now the reason for the 'arbitrary' enlargement of the east wing containing the portal in the inner court was not clear; its east side, with a width of 7.92 m, is considerably larger than the average 7.40 m of the court sides III–VII. There is a simple geometric explanation for this striking enlargement: due to the walls of the portal and the apparatus for the portcullis, the east wing is approximately 0.25 m greater in depth than the other wings of the citadel. Room I (the 'throne room') does not reflect this extra wall thickness; its dimensions are not influenced by it. At 6.42 m, the depth of this room even *exceeds* the average of all other rooms by 2 cm. The difference is compensated by displacement of the court wall in the direction of the inner court. A rough diagrammatic sketch (Fig. 234) shows that side I of the inner court thus becomes wider at the expense of the directly adjacent

162   *Design and Construction*

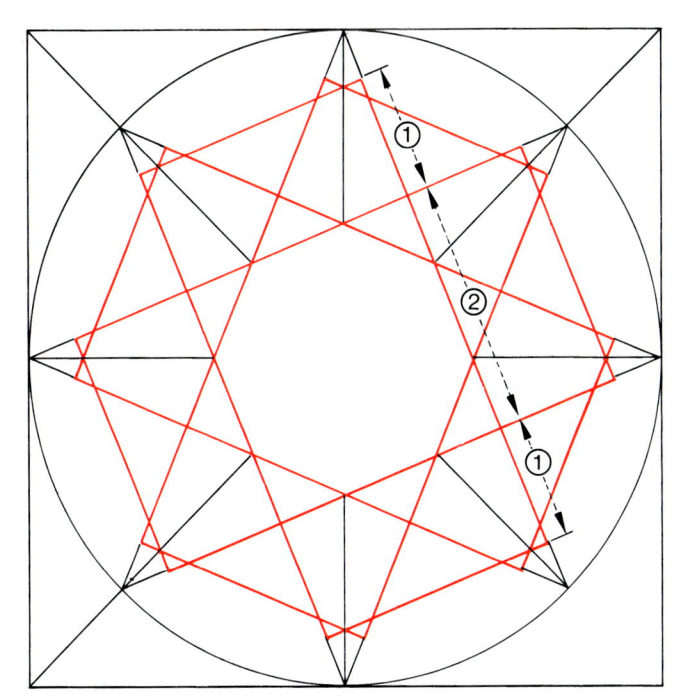

231 Step E: The octagon of the inner court (step D) consists of two intersecting squares. A rectangle half the size of the square is added to each of the eight square sides. The corners of these rectangles intersect with each other on the diameters and diagonals of the large square running through point 0 (black lines).

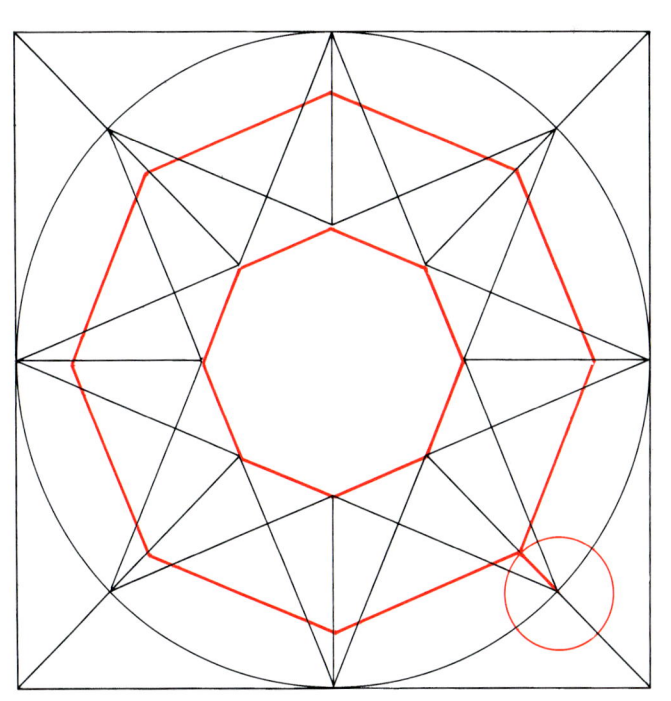

232 Step F: The distance of the points of intersection of the eight added rectangles on the diameters and diagonals of the large square, described under E, from the points of the eight-pointed star form the radius of the circle that circumscribes the octagons of the corner towers. The centers of these small circles are the eight points of the star.

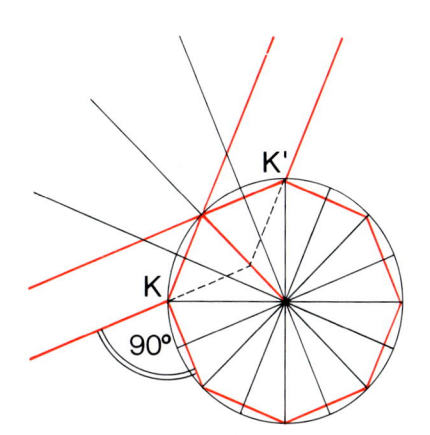

233 Step G: The construction of the plan for the eight octagonal corner towers. From points K and K' run the outer alignments of the outer walls of the main octagon. The inner alignments of these outer walls meet the invisible corner of each tower that is directed toward the center of the building.

court walls II and VIII, which thereby become narrower than the norm (see here in detail p. 173 ff.).

In a discussion of possible alternative methods of geometrically constructing the plan of Castel del Monte based on the eight-pointed star, Marcel Erné made another suggestion, which I designate here as (2); it, too, confirms the correctness of the geometric design plan proposed here (I am indebted to M. Erné also for the drawings in Figs. 236–243).

My proposal (1) is developed from the center and follows the path that showed me the geometric connections. It is completely in line with the construction of the eight-pointed star in the mausoleum of Humāyūn in Delhi (see p. 132, Fig. 192) and with the design pattern still used today by Moroccan craftsmen (see p. 132, Fig. 191). In this design the *towers* lie in the main directions of the compass rose. The orientation of the *portal wall* toward the east therefore requires a turn by 22.5°. We encountered the same problem in connection with the plan of San Vitale (see p. 136, Fig. 202).

The Geometric Construction and Measurements 163

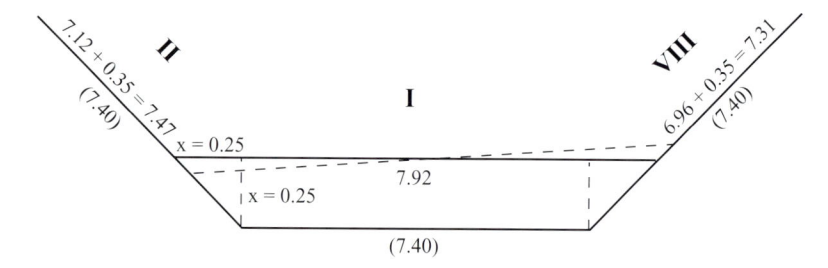

234 Castel del Monte, Apulia: Schematic representation of the irregular form of the inner court octagon on the east side of the portal wing (drawing bottom). Since the theoretical and the actual course of the inner wall of the octagon on side I are not exactly parallel, the wall of side II is somewhat longer than the wall of side VIII.

The advantage of the new solution is that the plan places the citadel walls, and thus also the portal in the east, in the compass rose from the outset (Fig. 235). The compass rose as the starting point recalls to mind the Portolan map of the *Carta Pisana* (see p. 146, Fig. 214).

Construction (2) begins with the making of a square grid. Such grids are familiar to us from the medieval architectural design process, particularly in Islamic architecture (see p. 141, Fig. 208).[236] Moreover, construction method (2) can be employed without the use of a compass.

Step A (2) (Fig. 236): Construction of the right angle (Fig. 228) as in step A (1).

Step B (2) (Fig. 237): Formation of a square divided into $4 \times 4 = 16$ squares with the length of the side h, which corresponds to the basic measurement (module), whose size

235 Castel del Monte, Apulia: Geometric ground plan with the eight wings in the main wind directions, i.e., rotated by 22.5° compared with the plan on p. 145, Fig. 213.

164   *Design and Construction*

236 Step A (2)   237 Step B (2)

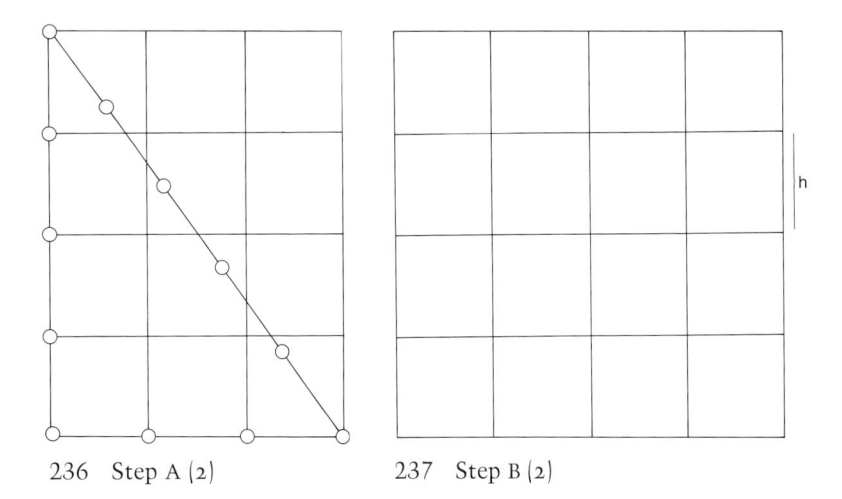

can be determined from the actual dimensions of the citadel (see p. 169ff.).

Step C (2) (Fig. 238): Formation of a second square through rotation of the first square by 45°, such that the corners of the new square fall on the extended main axes of the initial square. We are dealing with the repeatedly encountered basic figure of 'rotating' squares.

Step D (2) (Fig. 239): Construction of the eight-pointed star by extending the sides of the inner octagon (inner court) in both directions up to the eight points of intersection M_i, which mark the centers of the outer towers. This figure corresponds to the compass rose of the *Carta Pisana* (p. 146, Fig. 214), step D (2) to step D (1).

Step E (2) (Fig. 240): Construction of the peripheral octagon that touches the corner points C_i of the two concentric squares and at the same time forms the two outer sides of each octagonal tower. The corners of the new outer octagon are designated D_i.

Step F (2) (Fig. 241): Construction of the outer towers: E_i denotes the points of intersection of the middle octagon (corners B_i in Fig. 238) with the axes F_i and G_i. For the tower with midpoint M_i the sides are formed by the intersection of the following eight lines: E_1F_1, E_2G_1, C_1D_1, D_1G_2, F_1G_4, F_6G_1, E_2E_4, E_1E_7.

Step G (2) (Fig. 242): Construction of the outer walls by connecting neighboring towers.

The geometric ratios of size for the ground-plan design are shown in Fig. 243; the results are extremely interesting. The statements that follow are independent of the *method* of construction and therefore pertain both to method (1) and to method (2):

The Geometric Construction and Measurements   165

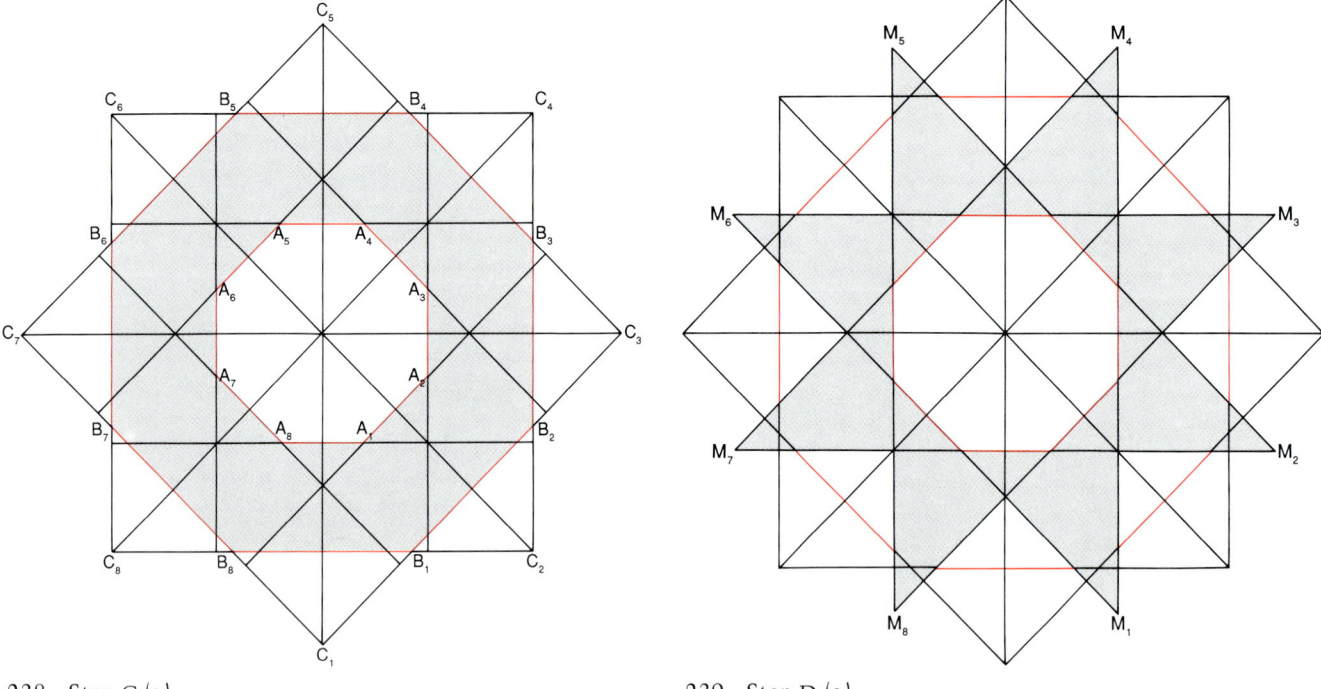

238 Step C (2)

239 Step D (2)

1. The sides of the octagon that forms the inner alignment of the outer wall are twice as long as the sides of the inner court octagon (denoted a in the drawing) — they represent the outer alignment of the court walls of the citadel.
2. The width of the eight outer towers theoretically also equals a, i.e., the length of the sides of the inner court (cf., however, Fig. 245).
3. The distance between two adjacent towers is $\sqrt{2}\,a$. The side length D_i of the peripheral octagon which is measured when a rope is stretched around the entire building (see step E [2]) is twice as large, namely $2\sqrt{2}\,a$.
4. The side length of the towers amounts to $b = a(\sqrt{2}-1)$. This size ratio appears frequently in all octagonal constructions and is mentioned repeatedly by M. S. Bulatov.
5. The outer wall thickness therefore amounts to $c = b/\sqrt{2} = a(1-1/\sqrt{2})$.
6. Mesh width h in the basic grid is at the same time the height in the triangle with base a, from which the inner court octagon is produced. The ratio of the two sides is $h = a(\sqrt{2}+1)/2$, i.e., approximately 6:5.

It is most interesting, and fully in accordance with the results of M. S. Bulatov's research, that almost all appa-

166   *Design and Construction*

240   Step E (2)

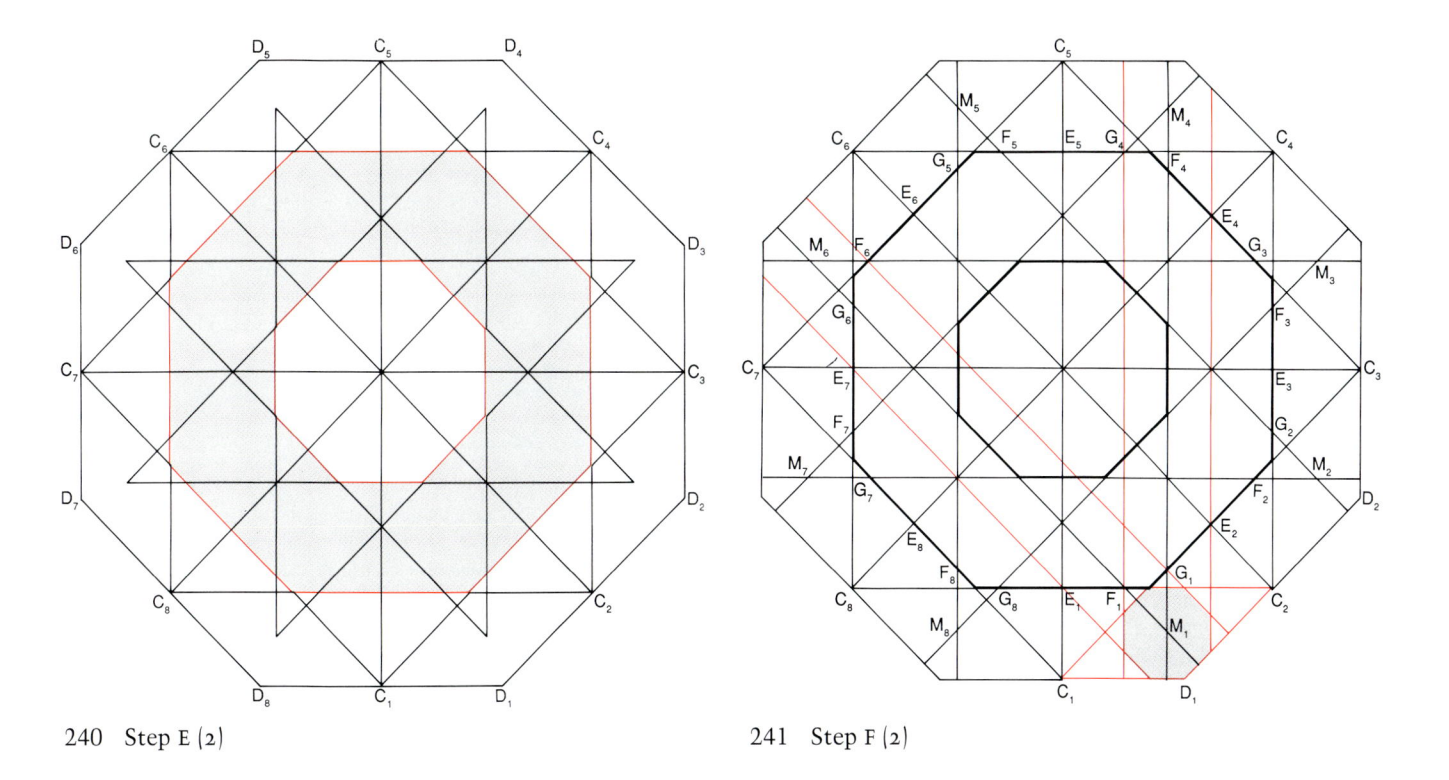

241   Step F (2)

rent distances can be expressed as powers of $\sqrt{2}$, multiplied by a, in the form $\sqrt{2}^n a$ with integral n ('geometric progression'!). The central importance of the quantity $\sqrt{2}$ to the plan of Castel del Monte has already been pointed out. Thus comes to light a decisive foundation for the consciously designed aesthetic.

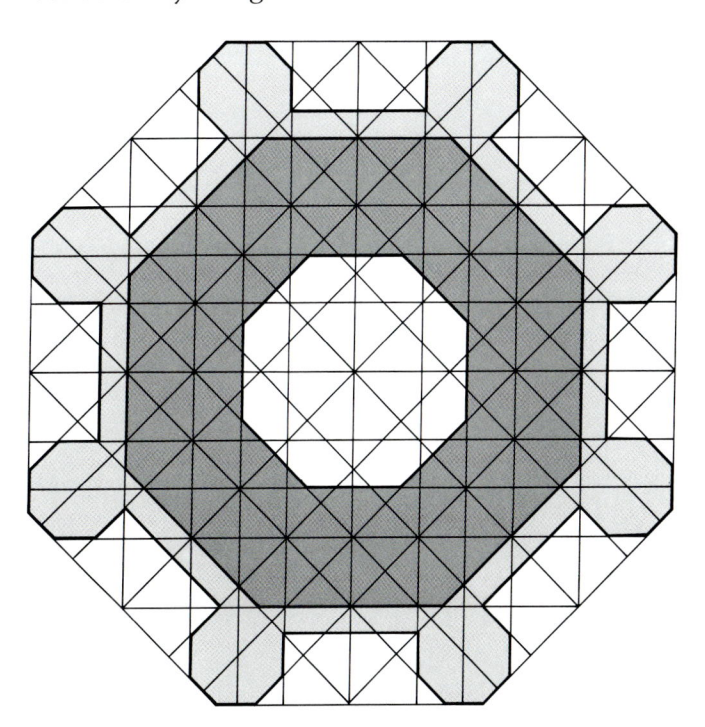

242   Step G (2)

The Geometric Construction and Measurements   167

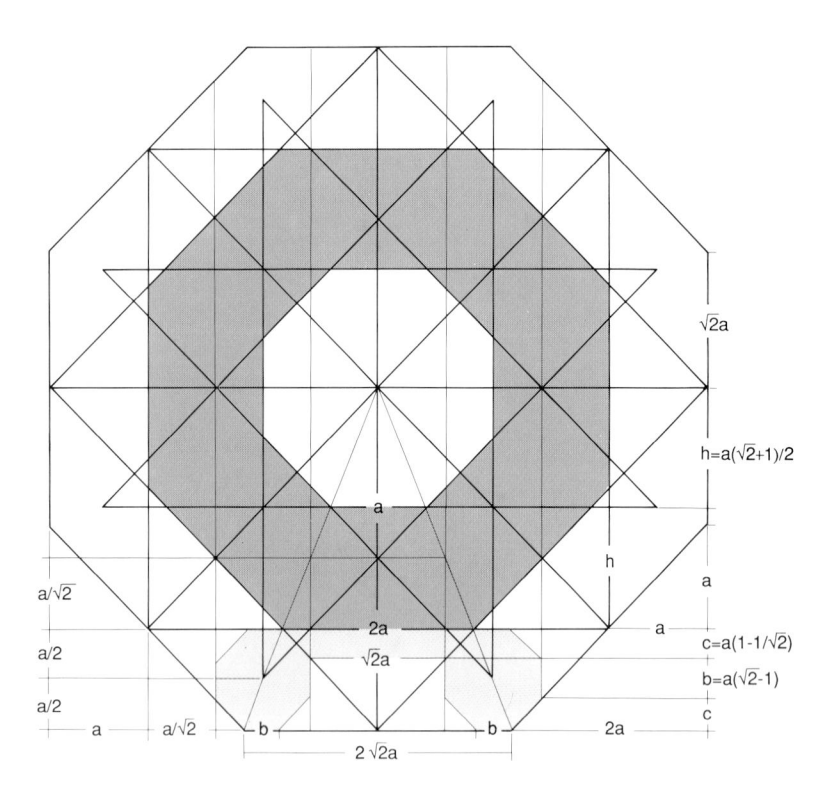

243 Geometric proportions of the ground plan construction.

According to this formula $\sqrt{2}^n a$, we obtain for

$n = 0$ the side length of the inner octagon — equal to the width of the towers

$n = 1$ the distance between the towers

$n = 2$ the side length of the inner octagon (inner alignment of the outer walls)

$n = 3$ the side length of the outer octagon (perimeter of the entire building).

The presentation of the size ratios in Fig. 245 takes, as already elaborated, the sides of the inner court, whose length is equal to the width of the eight outer towers, as the base quantity: 'a'. The other proportions can be obtained from this quantity a.

It would be just as correct, however, to choose the mesh width of the basic grid, designated as 'h' in Fig. 237, as the starting and reference quantity. The following relationships are obtained with the conversion based on the quantity h:

$a = 2(\sqrt{2}-1)h; \quad \sqrt{2}a = 2(2-\sqrt{2})h; \quad a/\sqrt{2} = (2-\sqrt{2})h$
$b = 2(\sqrt{2}-1)^2 h = 2(3-2\sqrt{2})h$
$c = \sqrt{2}(\sqrt{2}-1)^2 h = (3\sqrt{2}-4)h$

At first, both procedures appear to be equal. Quantity 'a' is impressively visible on the building as the width of the towers and the side length of the inner court. Quantity

168 *Design and Construction*

'h', on the other hand, has a more fundamental importance as a unit of measure of the underlying grid square (Fig. 237). This large grid square has a key function in the development of the plan, and it plays the decisive role in the transfer of the plan from the final drawing to the building site, in the 'laying out' with modules, or in the stretching of ropes.

Taking 'h' as the basic quantity, we obtain the following:
1. The distance between the outer alignment of the inner court and the inner alignment of the outer walls amounts to h.
2. The diameter of the inner court from mid wall to mid wall is $2\,h$.
3. The distance between tower centers is also $2\,h$.
4. The side length of the towers is $b = a^2/2\,h$, the thickness of the outer wall $c = a^2/2\sqrt{2}\,h$.[237]
5. The overall width of the citadel is $4\sqrt{2}\,h$.

The mathematical definitions given in Fig. 243 correspond to the geometric configuration. We will see that the architects made slight variations that did not affect the fundamental geometric concept — in the interest of accentuating the portal wing, for instance. The variations presuppose a regular basic plan from which they were able to deviate without detracting from the general impression (see the figure on p. 8).

If we pursue the question of the unit of measurement concealed in the two large intersecting squares that are the starting figures of the entire structure, we discover the Roman foot; equal to 29.57 cm (rounded up to 30 cm), it is contained 30 times in h (30 × 30 cm = 9.00 m). The side lengths of the large rotating basic squares are therefore about 120 feet each. The basis is a duodecimal system. In the Cistercian church of the abbey at Eberbach in the Rheingau, carefully studied by Hanno Hahn, one of the two principal measurements, the total cross-sectional width ('measurement II'), is 17.70 m ≅ 60 Roman feet. With a very small difference, this corresponds to the width of the inner court of Castel del Monte, which is 17.87 m. The total length of the nave, being approximately 71 m ≅ 240 Roman feet, measures exactly twice that of the total inside width of Castel del Monte, which is 36 m ≅ 120 Roman feet.[238] For the design plan of Castel del Monte, just as for the abbey church of Eberbach, it is

The Geometric Construction and Measurements 169

essential that measurements are not taken from outer alignment to outer alignment of the peripheral walls; the decisive foot measures of the buildings of that period are *inside* measures. Thus the diameter of the inner court measures two grid widths ($= 2h$) $\cong$ 60 Roman feet. The grid measure $h \cong 30$ Roman feet is the starting measure for all further distances pertaining to the citadel. All further measures are sequential measures that can be determined from the various already explained geometric relationships (Fig. 243). It is interesting to note that the middle room of the basilica at Fano, built by Vitruvius Pollio,[239] had an inside length of 120 Roman feet and an inside width of 60 Roman feet![240] The Cistercians thus followed the ancient Roman tradition of measurement.

Let us now compare the geometrically determined dimensions and their relationships to one another with those measured on the building itself. Mr. W. Schirmer was kind enough to apprise me of a number of data from the work of his group, even prior to his own publication and interpretation of the results of the measurements made. They appear here in Fig. 244.

The distance between two opposite court walls in the north-south direction is taken to be 17.86 m. Further measurements (see Fig. 244) have shown 17.85 m and 17.88 m. The reason for the deviation of the fourth measurement in the east-west direction (17.63 m) is given on p. 173f. We are dealing with the measure of the length of two grid squares, i.e., $2h = 18$ m $\cong$ 60 Roman feet. The difference from the average measurement of 17.87 m is negligible.

To give a better general idea, these and the following measurement comparisons have been entered in Fig. 245, the calculated ones in black and those measured in red.

Let us now move further outward and attempt to verify the distance between the outer alignment of the court wall and the inner alignment of the outer wall, which corresponds to the quantity 'h', according to our premises. The thickness of the inner wall, which was not determined by a geometric relationship and could be fixed arbitrarily, is 2.37 m in four wings, 2.36 m in two, 2.32 m in one, and 2.38 m in the last, giving an average of 2.36 m. To this must be added the likewise very uniform room depth of 6.40 m on the lower floor. The result is 8.76 m, a difference of 24 cm from 9 m. According to information from Prof. Schirmer, the room depth was measured

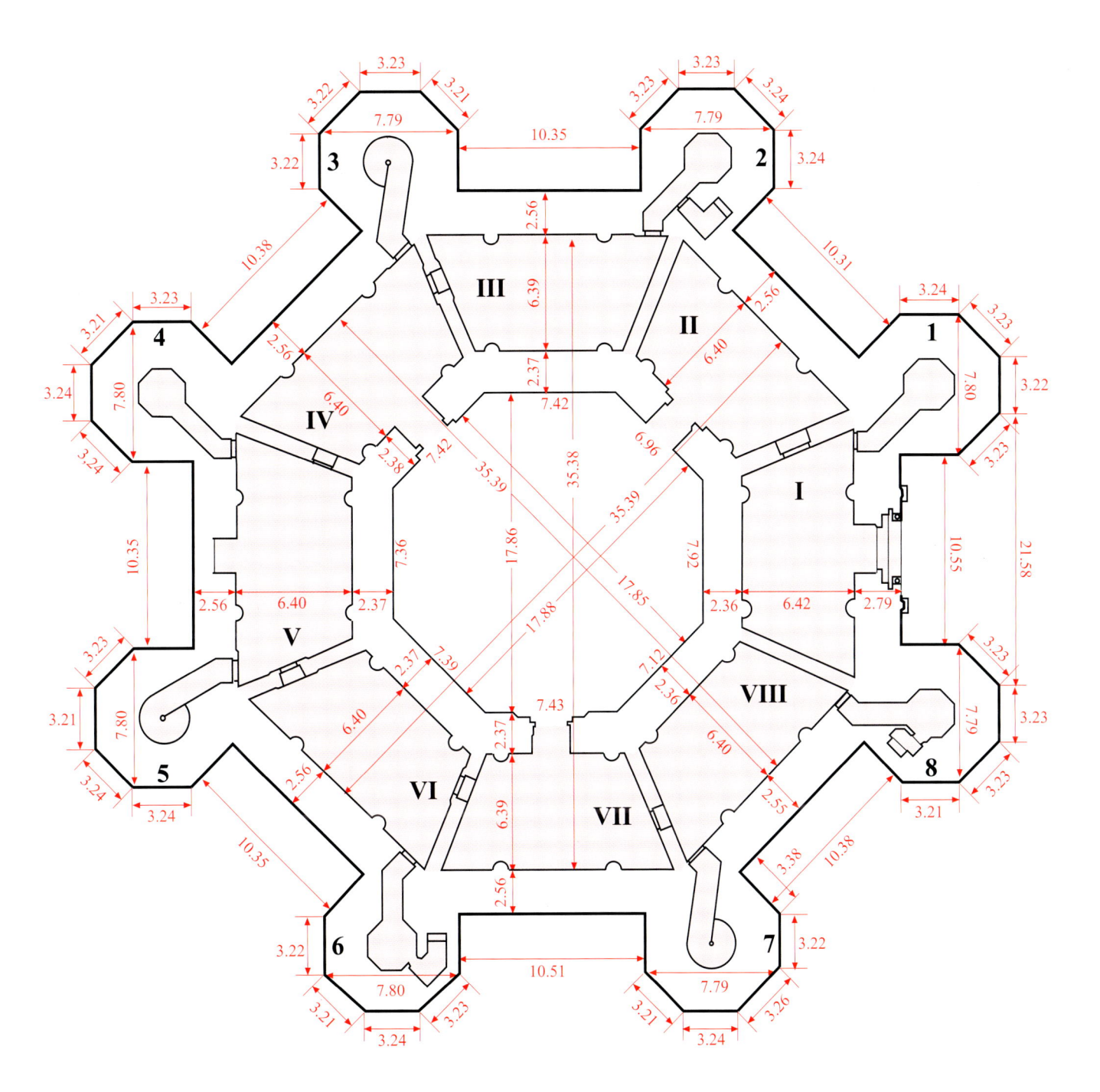

244 Castel del Monte, Apulia: Ground plan (lower floor) with results of measurements (in meters) done by the working group of W. Schirmer.

between the surfaces of the wall incrustations, which are estimated to be about 15 cm each. But for 2 cm, the inner wall thickness of 2.36 m corresponds to 8 Roman feet, and the interior room depth of 6.40 m + 15 cm = 6.55 m would amount to 22 Roman feet, resulting together in the basic quantity $h \cong 30$ Roman feet.

Let us now calculate the dimensions of the inner court sides, i.e., of the inner court octagon, from the ratios of the basic geometric plan. An inner court side is accor-

The Geometric Construction and Measurements 171

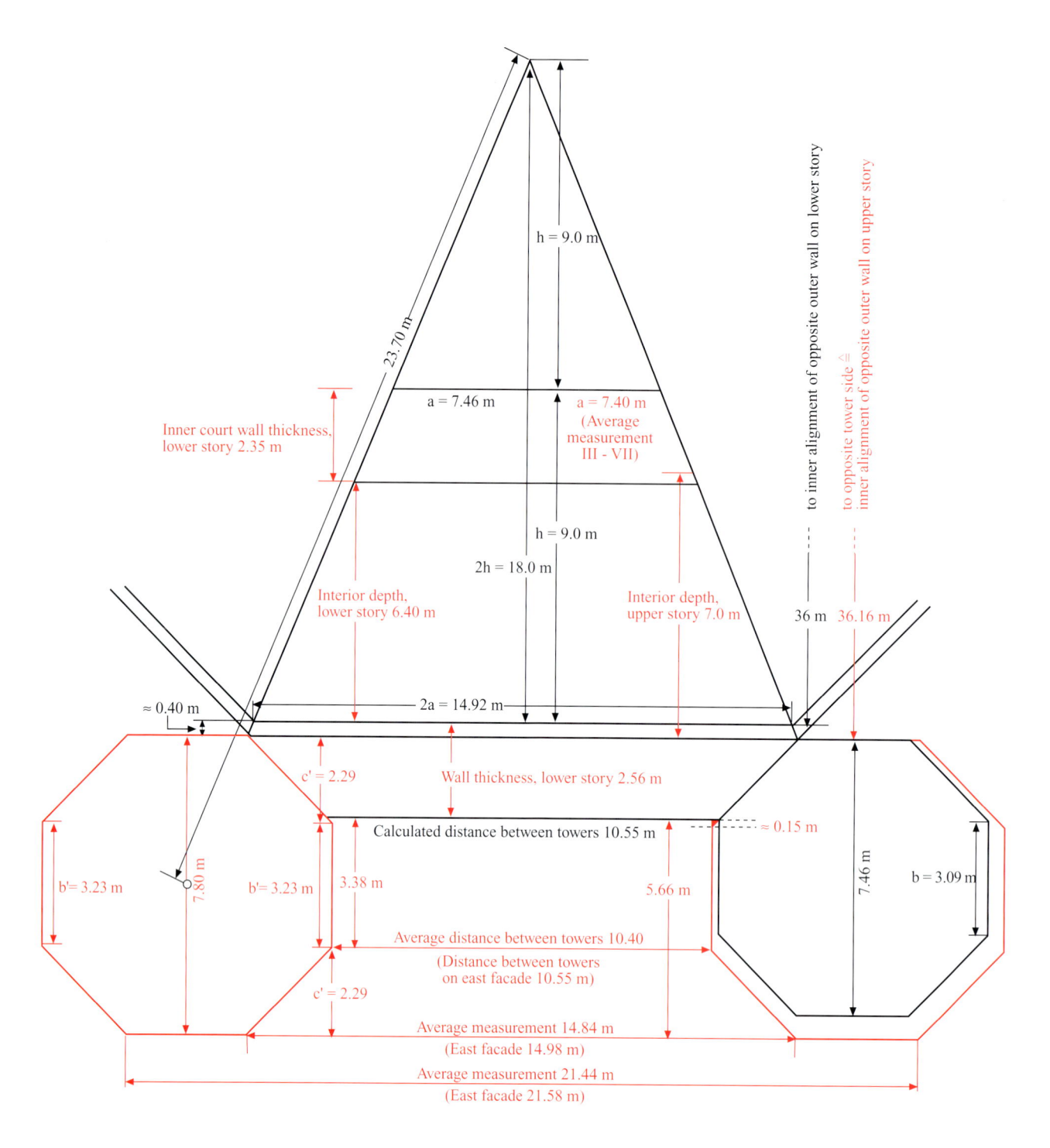

dingly $a = h/(\sqrt{2}+1)/2$, i.e., 9 m/1.207 = 7.46 m long. In a comparison with the measured dimensions (Fig. 245) the exceptional length of 7.92 m of the portal wing in the east must be disregarded for the time being, as must the below-average dimensions of the adjacent wings II and VIII. The average of the remaining five sides, of 7.42 m, 7.42 m, 7.36 m, 7.39 m, and 7.43 m, is 7.40 m, which differs by only 6 cm from the calculated length.

245 Outline of the system of measurement. The calculated dimensions are shown in black, those measured in red.

There is a simple geometric explanation for the marked deviations of the dimensions of the three eastern sides of the inner court. The eastward-directed portal side is more powerfully constructed than all other wings. The magnitude of its importance is expressed in more imposing dimensions, in part practically forced, for instance, by the accommodation of the splendid portal with the portcullis. However, the measurement increases remained within a scale that did not detract from the overall impression made by the regularity of the structure. The outer wall on the eastern side, at 2.79 m, is 0.23 m thicker than all the other sides (2.56 m). Room I, the throne room, does not compensate for this greater wall thickness — it is even 2 cm deeper than the other rooms, meaning that the added dimensions jut further into the inner court over its normally sized wall. In this way the regular form of the inner court octagon was changed, as is apparent from Fig. 244. If one assumes that the eastern side of the inner octagon extends by 25 cm into the court area, the result is a broadening of this court side by 2 × 25 cm = 50 cm, since an equilateral triangle with legs of 25 cm is formed in each corner. If these 50 cm are added to the normal 7.40 m width of the inner court wall, the result is 7.90 m — almost identical to the measured wall width of 7.92 m! The correspondingly shortened inner walls II and VIII are thus, according to Fig. 234, $y = x \times \sqrt{2}$ or $25 \text{ cm} \times \sqrt{2}$ = 35.36 cm. If we add 35.4 cm to the length of side II, measured at 6.96 m, we get 7.31 m, i.e., 9 cm *less* than the average measurement of 7.40 m. For side VIII we get 7.47 m, about 7 cm *more* than the norm of the other sides. The fact that the actual dimension in one case (side II) falls short of, and in the other case (side VIII) exceeds, the calculated dimension suggests that court side I, which juts out, should run parallel to the geometrically fixed alignment of the court side but actually deviates from this strict parallelism (see p. 164, Fig. 234, broken line).

It stands to reason that these displacements for the sake of the eastern portal had to interfere slightly with the geometric exactness of the overall design — for example, the radial connection of the citadel center with the centers of the outer towers. The unaltered course of the outer wall remains remarkable. This is further proof of the hypothesis already put here that the belt of the large citadel wall, based on the figure of the crossed squares, represents the principal reference system, determined by the quantity h.

A further area of certain deviations in measurement are the towers. Although the towers themselves display a consistent width of 7.79/7.80 m, the distance between them varies from wing to wing. Thus the inner distance between the tower walls of the east wing (I) is 10.55 m, while the corresponding distance for the adjacent northeast wing II is only 10.31 m. The distance between the towers of the adjacent southeast wing VIII is 10.38 m (Fig. 244). What we discovered with regard to the eastern inner court wall is repeated here: The east wing is accentuated not only by the extended depth but also by the increased width; the enlargement is absorbed and balanced by each of the adjacent wings II and VIII.

The distances between the remaining towers are 10.35 m (III), 10.38 m (IV), 10.35 m (V), 10.35 m (VI), and 10.51 m (VII). The last of these is an exception, as all other distances — save that of the portal wing — are in the range of 10.35 m. The overall average is 10.40 m as against a calculated distance of $\sqrt{2} \times a = 10.46$ m!

According to the geometric plan drawing (Fig. 245), the two inward-directed sides of each octagonal tower, which are either not visible or visible only on the roof, form the extension of the inner alignments of the outer walls. Consequently, the inner octagon sides that lie exactly opposite — and appear only on the roof — should be the same distance from one another as the opposite sides of the large basic squares, namely $4h = 35$ m $\cong$ 120 Roman feet. Measurements made with tape on the roof, where the inward-directed sides of the towers become visible, showed a distance of 36.16 m. With a slight deviation, this agrees with what is expected on the basis of the plan.

If we add together the individual distances between the inner alignments of the outer walls on the ground floor we have 17.87 m court diameter, plus 2×2.36 m $=$ 4.72 m courtyard walls, plus 2×6.40 m $= 12.80$ m inside depth of the rooms, for a total of 35.39 m. This is a difference of 0.77 m from 36.16 m, i.e., ca. 39 cm on each side. If we subtract this 0.39 m from the outer wall thickness of 2.56 m, the remainder is 2.17 m. This should correspond to the dimension for c in our basic plan — $c = a(1 - 1/\sqrt{2}) = 7.40$ m $\times 0.293 = 2.17$ m — which is the case. The measurements are no longer identical, however, if we do not calculate the quantity *c* from our determined value $a = 7.40$ m but take as our starting point the measured tower width $a' = 7.80$ m. We then get $c' = 2.29$ m.

6 Spandrel Formation

Castel del Monte is a construction with an amazingly exact system of measurement concerning the court octagon, the large outer wall, and the towers. We will go into

246 Castel del Monte, Apulia:
East side with the protruding alignment of the portal, which continues up to the upper edge of the east-wing wall. To avoid recesses of the upper rim, the portal alignment is carried through to the adjacent tower sides by means of small arches.

175

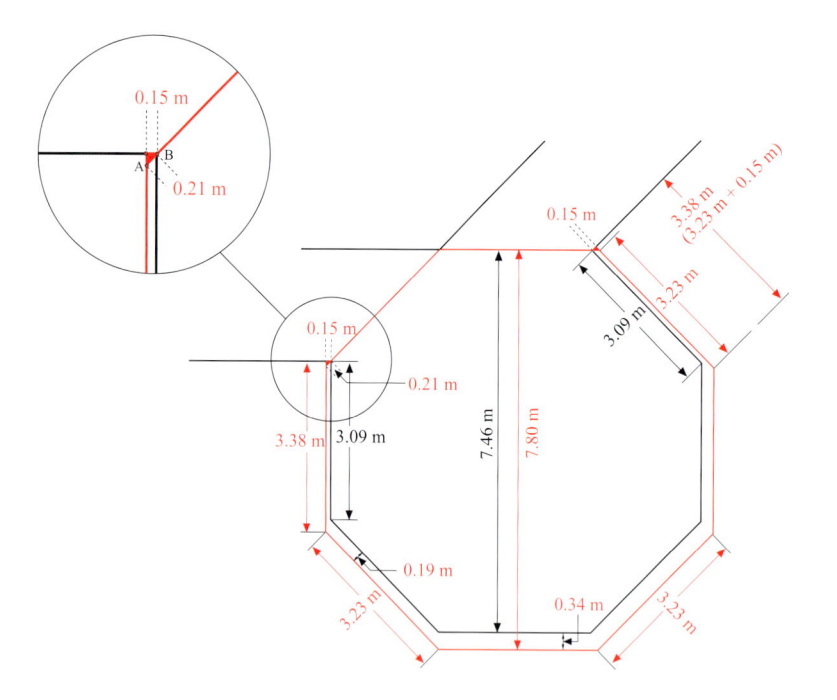

247 Castel del Monte, Apulia: Schematic representation of the proportions in the 'spandrel zone' (see also Figs. 248, 249).

this in detail in chapter VI (p. 191 ff., Fig. 257). Two complexes that deviate from the expected dimensions have already been discussed above (p. 162 and Fig. 234).

The sides of the eight corner towers are consistently ca. 3.22/3.23 m wide — an astonishing conformity for a total of 48 sides directed outward. Those inward-directed sides that meet the wall of the main octagon at an angle of 90° are consistently about 15 cm wider, however (ca. 3.38 m). This difference from the normal width of the side walls has generally been overlooked in descriptions of the plan and probably attributed to the allowable deviation of 'medieval' buildings (only W. Schirmer shows the longer sides clearly in his plans,[241] but without explanation). However, if one observes the area in which the longer side wall meets the wall of the main octagon, one soon becomes aware that there is a clearly manifest reason for the broadening of these inner walls by ca. 15 cm: The wall that runs toward the main octagon ends in reality after the normal width of ca. 3.22/3.23 m and then continues — like a normal octagon side — at an angle of 45°; it meets the wall of the main octagon only after about 21 cm (see Fig. 247, especially the detail in the circle). The entire side wall does not follow the 45° bend at A, but continues to run perpendicularly toward the main wall. The result is, as stated, the greater width of this octagon wall of 3.23 + 0.15 = 3.38 m.

176 *Design and Construction*

248   Castel del Monte, Apulia: View of the angle between the south wing of the main octagon wall and tower 6. A space between the inner edge of the tower and the main wall is clearly visible.

249   Castel del Monte, Apulia: Example of a 'wall spandrel', found between the upper edge of the outer wall alignment of the main octagon and the inner edge of the east side of the sixth tower. See also Fig. 250, 20 VI L.

Moreover, this extension of the wall by 15 cm, along with the short run of the regular octagon wall (to B) and the section of the main octagon wall, produces a spandrel in the form of an isosceles triangle with legs of ca. 15 cm each and a base of ca. 21 cm.

It is astounding that these spandrels occur on all towers (twice each). They are recognizable, however, only at the roof level of the towers. I have photographed each one individually and arranged them together in Fig. 250. It is conspicuous, and indicative of the exactness of the measurement system, that all 16 visible spandrels exist not only in the expected form but also, on average, in the same shape and size.

Is there a plausible reason for this spandrel formation? In Fig. 245 I have drawn in black the dimensions that may be 'theoretically' expected based on the geometric concept of the plan, in red the dimensions actually measured.[242]

If we now construct the regular tower octagon starting not from corner A but from the spandrel corner B (see Fig. 247), we find, to our surprise, that the resulting regular octagon is not 7.80 m wide, according to the measured dimension, but only 7.46 m, corresponding exactly to the theoretical width of the towers calculated by Prof. M. Erné!

We must therefore conclude that, for reasons unknown to us, the width of the towers at the start of construction was increased, in fact, to the same degree for every tower. Since nothing was altered in the system of

Spandrel Formation   177

13 I L

12 II R

11 II L

8 IV R

7 IV L

250 Photographs of the spandrels on all towers where they are visible, excepting sides I and VIII (portal side). The positions of the individual towers are noted in the accompanying drawing. It is striking that of the 14 spandrels 12 are clearly visible and have on average the same shape and size!

21 VI R

20 VI L

178   *Design and Construction*

III R　　　　9 III L

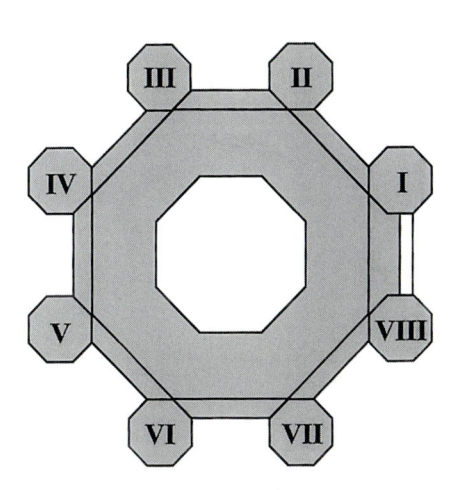

V R　　　　22 V L

VII R　　　　18 VII L　　　　17 VIII R

Spandrel Formation

measurement as a whole, the inevitable result was the spandrels we have observed, of which, amazingly, no note has been made in any picture of the building up to this point (not even by W. Schirmer[243]). It should also be noted that none of the towers shows any evidence that "by means of continual correction during the building process the accuracy of the basic geometric form" was improved. The dimensions of the towers overall and in detail — as the picture of the entire citadel clearly shows — display instead an astonishing, continuous faithfulness of dimension. We will return to the subject of this internal unity of the exact system of measurement in chapter VI (p. 191 ff.).

In what way, however, does the large octagon with the dimensional foundations that we have determined, which stem from the basic quantity h and can be traced back to the classical figure of the crossed squares, relate to the layout of the ground floor? The inner alignments of the outer walls there run about 0.40 m further inward than the line that is defined by the two 'inner' octagon sides of the towers. It must be assumed, in principle, that the binding layout for the practical construction plan was the ground floor. It is also conceivable that in the plan, elaborated with great precision probably by Cistercian builders, the layout of the towers was included in the considerations from the start. It is noteworthy in this connection that on the upper floor with its thinner walls the inside depth of the rooms measures 7 m, in contrast to 6.40 m on the ground floor. This means that the inner alignment of the outer wall coincides almost exactly with the large octagon, which is coincident in turn with the inner sides of the octagonal towers, if one estimates that the outer walls taper by 30–40 cm (exact measurements of the wall thickness on the upper floor are not yet available).

We can only speculate, of course, as to how the geometric design plan was actually transferred to the building site. Both the stretching of ropes following determination of a certain fundamental direction, as has already been described here, and the transfer of enlarged units of measure (modules) to the building site from a design plan covered with a grid are imaginable. A combination of both methods is also possible.

Independent of how accurately the design was transferred to the building site, it is inconceivable that the builders could have managed without a basic geometric plan of the sort that has been reconstructed here. After all, every section and construction unit of Castel del Monte consists in strict geometric configurations, all of them related to one another. It is impossible to imagine how such a building could have been completed in practice without a drawing of the plan that connects the visible geometric elements with one another. I have repeatedly called attention to the method of the rotating or crossed squares for constructing an octagon; in the case of Castel del Monte, as well, it appears to have been the fixed point of departure for the transfer of all measurements. If Bulatov finds large constructional deviations in the realization of geometric concepts by experienced Arab builders, how can we expect anything different of Castel del Monte? These deviations are nevertheless amazingly small.

182

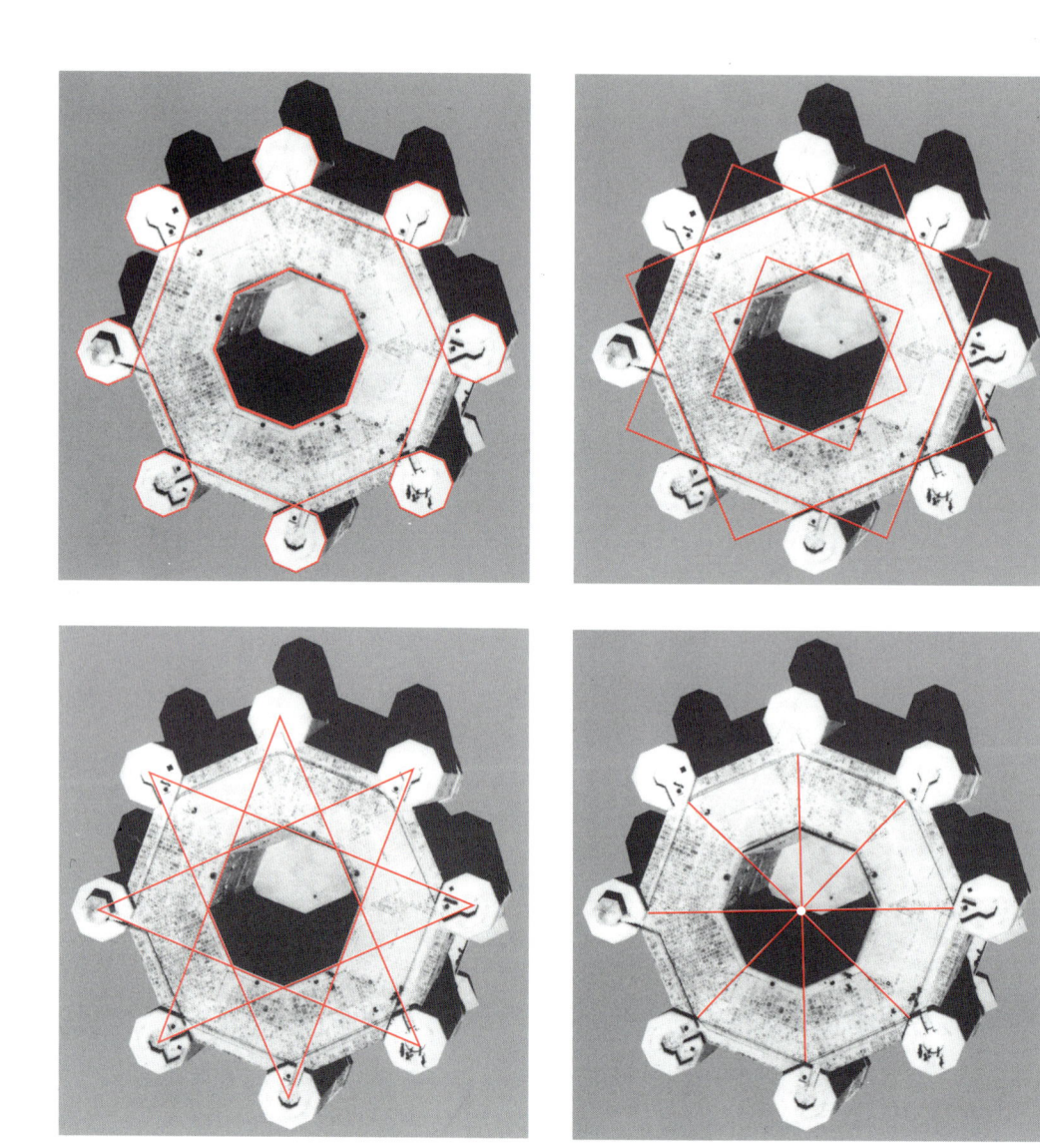

V

The Geometric System and Its Realization

The bird's-eye view of Castel del Monte reveals the ground plan. Stereometrically, we are looking at a so-called monohedron, a 'single surface', in contrast, for instance, to a hexahedron with six identical surfaces (a cube). A monohedron is a solid body which — independent of its height — is defined only by its horizontal expanse. Matthias Untermann[244] therefore correctly concludes that a centrally planned building is recognizable at a glance from its ground plan. A horizontal section through the upper story is practically identical to a section through the lower story; constructional details — for example, pertaining to the foundation or the wall thickness — or interior divisions alter nothing in principle.

The planimetric photo shows that the midpoints of the octagonal towers coincide with the points of intersection of the extended sides of the inner court (p. 159, Fig. 226). This relationship is so clear that it requires no further proof. It is so unambiguous that it cannot be the product of chance.

All of the four possible subgroups of the eight-sided polygon appear in the plan of Castel del Monte: the simple octagon, the simple eight-pointed star formed from two intersecting squares, the pointed star polygon, and the four symmetrical axes. The star polygon is also known as the 'Islamic eight-pointed star' because it occurs particularly frequently in Islamic ornamentation. It is the most striking form of the eight-sided polygon. The fact that all four described subgroups of this eight-sided polygon are incorporated in the building proves that the architect considered the symmetries systematically (Fig. 251 a-d).

I have already explained in detail the geometric construction of the eight-sided polygon and of the star poly-

251 a–d
Seen from a bird's-eye view, the four forms of the octagram in the citadel are clearly recognizable.

gon (eight-pointed star). On the basis of the calculated geometric connections, the relation between the three 'similar' (homeomorphic) octagons can be defined thus: The distance between the inner alignments of the outer walls is twice that between opposite walls of the inner court. The corresponding distance between opposite sides of the octagonal towers at the corners of the large polygon has the value $(\sqrt{2}-1)$. These relations also exist between side 2a of the outer octagon, side a of the inner court, and side b of the corner tower = $a(\sqrt{2}-1)$. The proportional relationship of the three octagons which determine the ground plan is thus $2:1:(\sqrt{2}-1)$. This formula defines the geometric relationship of all units of horizontal measurement in the building with one another. The dimensions measured at the building confirm this geometrically determined ground plan, with the exception of several intentional deviations which have already been discussed (see p. 162 and Fig. 234).

I am not aware of any European fortress which has such clear regularity in its ground plan design.[245] No comparable geometric construction of this type can be found in the lodge book of Frederick II's contemporary, Villard de Honnecourt![246] Certain regularities of a related kind

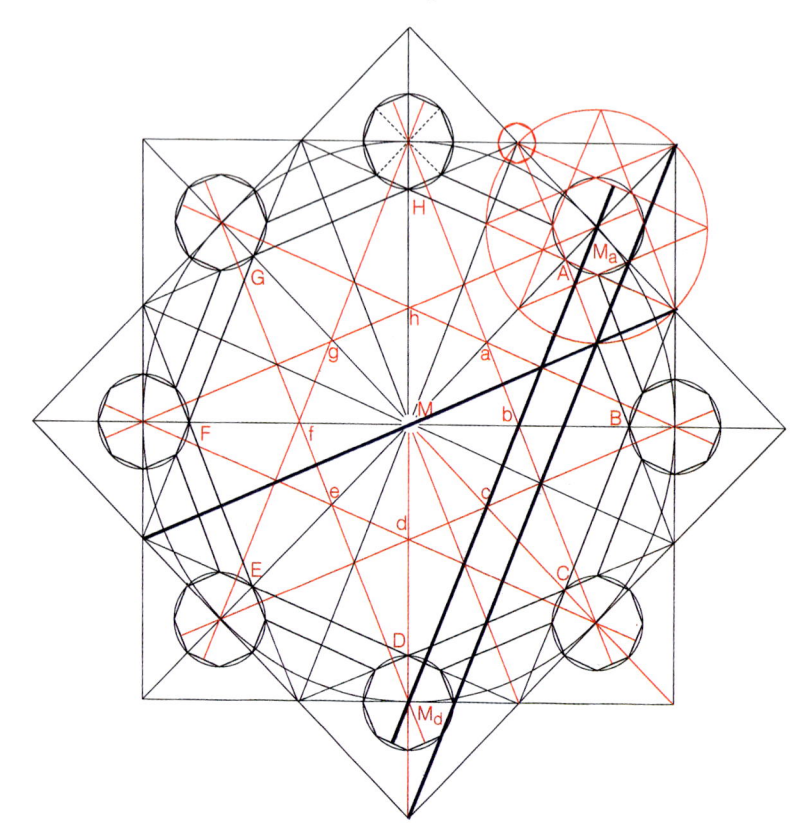

252 The "configuration with an internal aesthetic" (M. Koecher) is characterized by the frequency of collinearity, i.e., linear systems with numerous common points of intersection.

184 *The Geometric System and Its Realization*

are found, however, in centrally planned sacred buildings of the Middle Ages which ultimately derive from classical models.[247] M. Bulatov[248] and others have shown how the symmetrical architecture of the Islamic world draws on related Byzantine sources.

The geometric regularity of the plan of Castel del Monte is expressed primarily by a large number of symmetrical relationships. Max Koecher[249] has defined this abundance of symmetry as a "configuration with an internal aesthetic." It is characterized by the frequency of so-called collinearity, i.e., linear systems with numerous common points of intersection of various lines (Fig. 252).

There is proof for the reliability of our geometric definitions: the building can be constructed by computer simulation using the formula determined for the ground plan, $2:1:(\sqrt{2}-1)$, without the input of any ground-plan measurement data. This has kindly been done by Susanne Krömker of the Interdisciplinary Center for Scientific Calculation at the University of Heidelberg (Director: Prof. Dr. W. Jäger). The height corresponds to the relationship between the overall width and height of the existing building (Fig. 253).

We do not know how the design was transferred to the construction site — with modules, posts, or guide ropes.[250] The manifest existence of a geometrically constructed ground plan with its precisely measurable discrepancy from the actual building shows clearly that an exact plan *must* have been the basis for this geometrically complex structure, and that 'reality' deviated from this plan because of relatively minor, deliberate corrections.[251]

The above-mentioned primary existence of the outer alignment of the large basic octagon implies, along with numerous other considerations, that the building of this large main octagon was the first thing to be undertaken — irrespective of the sequence in which the geometric ground plan was designed.

Still to be discussed is the question of what ideas served as the basis of the building plan, which presents the octagon alone in so many guises. I have been able to demonstrate[252] the long tradition of the octagon, starting from the Hellenistic-Byzantine period and above all in Indo-Arab ornamental art and architecture. The eight-sided polygon, especially in the form of the star polygon, has remained unaltered through the centuries.

In the Codex Dioscurides of Vienna, dating back to the beginning of the sixth century (prior to 512 A.D.), the simple eight-pointed star formed from two intersecting squares appears, with the typical over- and undercrossing of the sides[253] (Fig. 166c). The motif is rooted in the Greek-Hellenistic tradition of eight-pointed star symbolism. An example from Vienne-Isère dating back to the second century A.D. (Fig. 166a) shows how the motif appeared in Roman times.[254]

Let us return to the basic formula elaborated above for the ground plan of Castel del Monte: $2 : 1 : (\sqrt{2} - 1)$, which defines the proportional relationship between the outer octagon, the inner octagon, and the corner tower octagons. How close and geometrically exact these relationships are becomes clear if one repeats the procedure of extending the inner sides of the court octagon to the midpoints of the corner tower octagons. New eight-pointed stars are the result, with points that touch one another (Fig. 254).

253 A reconstruction of the citadel by computer simulation based on the formula $2 : 1 : (\sqrt{2} - 1)$ proves the correctness of the formula. The illustration shows the continuation of the fundamental construction ad infinitum — a fractal with iterations. Two times the unit a_1 (the diameter of the inner court) was taken as the total height of the building — which has not been otherwise established — in this computer representation. Each tower is thus as high as its distance from the building center.

186 *The Geometric System and Its Realization*

254 The exactness of the geometric relationships becomes apparent if the procedure of extending the interior sides of the courtyard octagon to the midpoints of the corner tower octagons is repeated. New eight-pointed stars are produced, with points that touch one another (cf. Fig. 255). This diagram is a vector graph (produced by Susanne Krömker) and as a computer file can easily be scaled.

The succession of connected eight-pointed stars in the ornamental pattern above the tomb chamber in the mausoleum of Humāyūn in Delhi (Fig. 255), already mentioned, shows that such geometric figures are not unusual in the Indo-Arab world. There the rows meet with one point of each eight-pointed star. If the axis of this row of eight-pointed stars is rotated eight times by 45°, so as to produce a circle of 8 × 45° = 360°, the result is the figure

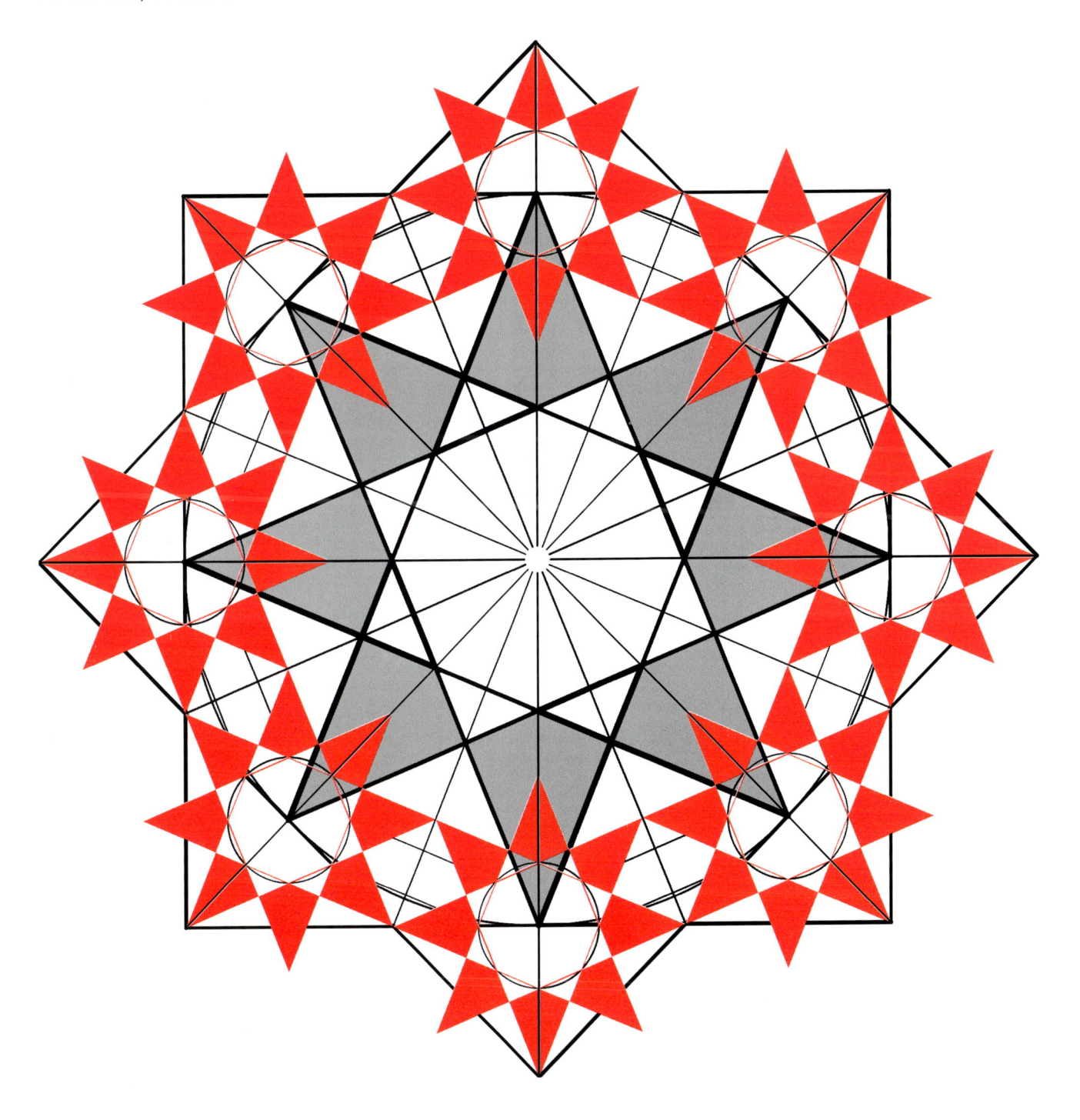

The Geometric System and Its Realization 187

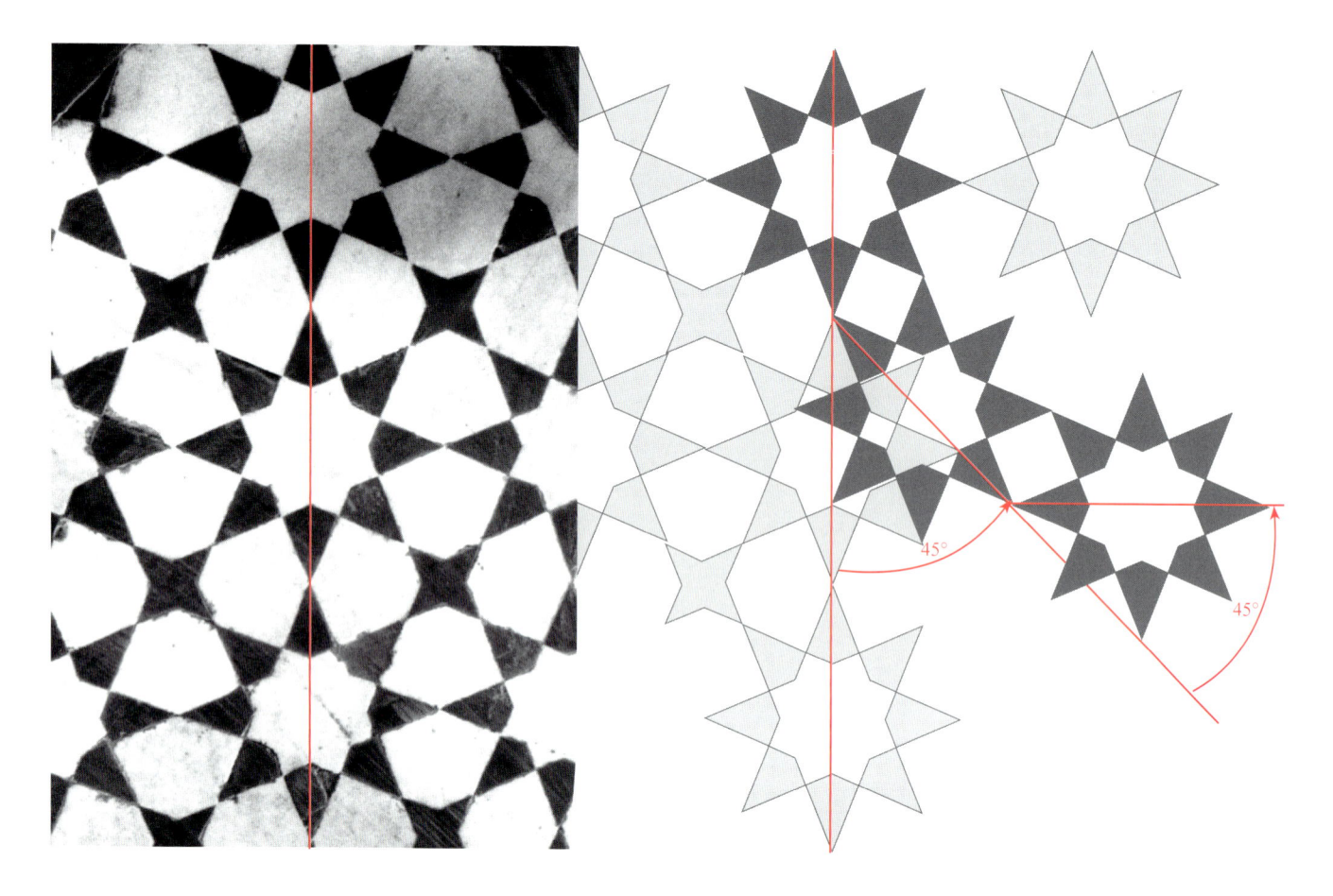

from Castel del Monte of eight-pointed stars meeting each other with *two* points each (Fig. 254). This represents an expansion and deepening of the octagonal motif.

This is further evidence for the correctness of the reconstructed geometric form of the building design. The mathematician Max Koecher has taken this continuation of the basic design construction to infinity — it is known as an infinite iteration.[255] The result is a 'fractal' (see Fig. 253), after the studies by Mandelbrod. The question of whether or not the creator of the building design was aware of this idea of continuing the process is irrelevant to our considerations. It simply serves as a proof of the geometrically exact nature of the design. It is interesting, however, that a fractal appears as a decoration on the floor of hall VIII on the first story (see Figs. 150, 151). Here it is essential to note that, in the case of the eight corner towers, we are dealing not with a more or less decorative extension of the basic octagon with its octagonal inner court, but with a systematic geometric continuation of the basic design concept following the formula we have determined: $2 : 1 : (\sqrt{2}-1)$.

255 Delhi: Mausoleum of Humāyūn. Eight-pointed star motif in the burial chamber. If the three eight-pointed stars standing above one another (left) and meeting one another at one point each are deflected from the vertical by 45°, the result is as shown on the right, i.e., the ground plan of Castel del Monte, in which the tower octagons are expanded to eight-pointed stars that meet one another at two points each (cf. Fig. 254).

188   *The Geometric System and Its Realization*

It is tempting to interpret the consistent extension of the octagonal motif in the eight corner towers as being a superelevation and thus an augmentation of the symbolic power of the octagon. This octagon, in particular the compass rose, radiating in all 'heavenly' directions, is a symbol of infinity and eternity. It has therefore also become the symbol of the worldwide power of empire.[256]

As can be proved, both the geometric form of the octagon in its variations and its symbolic power are of great importance to the Islamic environment of Frederick II. In contrast, a ground plan like that of Castel del Monte cannot be derived from the lodge book of his contemporary Villard de Honnecourt.[257] The actual process of building probably was attended to by the Cistercian order, committed to central European Gothic, but the complex geometric *design* with its configuration of multiple symmetry stems from another world.

Furthermore, at Castel del Monte, on the basis of what we have learned in defining the geometric ground plan and its practical realization, the conspicuous separation of *building design* and *building execution* becomes impressively evident — an observation that is important for architectural history.

256 Castel del Monte in the early morning sun; view from the road leading from Andria

VI

Provenance of the Plan

Visible from a long distance, the striking symmetry of Castel del Monte with its eight towers of equal size, its walls running strictly parallel to each other and to the main exterior wall of the citadel (see Fig. 256), is one of the chief characteristics of this architecture. How exactly parallel the surfaces really are is made clear by the identical shades of gray, i.e., the same angle of incidence of light on parallel surfaces.[258] From the very beginning it seemed to me that this strict symmetry, connected with the essence of geometry, was attributable not so much to the area of central Europe as to that of Islam, whose scientific achievements Frederick II so greatly admired. Studies and research in mathematics, particularly in geometry where Islam is concerned, which began at the latest in the eighth century, have their foundations in Greek tradition, especially that of Euclid.[259]

The geometric relationships mentioned above correspond to the dimensional accuracy of the building components. Thus with the distance between the inner surfaces of the outer walls on the ground floor measuring between 35.37 and 35.39 m, the opposite outer walls of the citadel on three axes differ in length by only 2 cm. The fourth, east-west, axis is an exception due to the portal construction.

In the drawing of the plan by W. Schirmer (see p. 171, Fig. 244), the distances between two pairs of opposite towers (1–6, 4–7) measure 36.18 m and 36.16 m, i.e., also a deviation of only 2 cm.

The eight towers differ by only 1 cm in width in each case: four towers are 7.80 m wide each, four are 7.79 m (Fig. 257). Also remarkable is the strict uniformity of dimensions as regards the width of the ground-floor interior rooms from outer to inner wall: in five cases it is

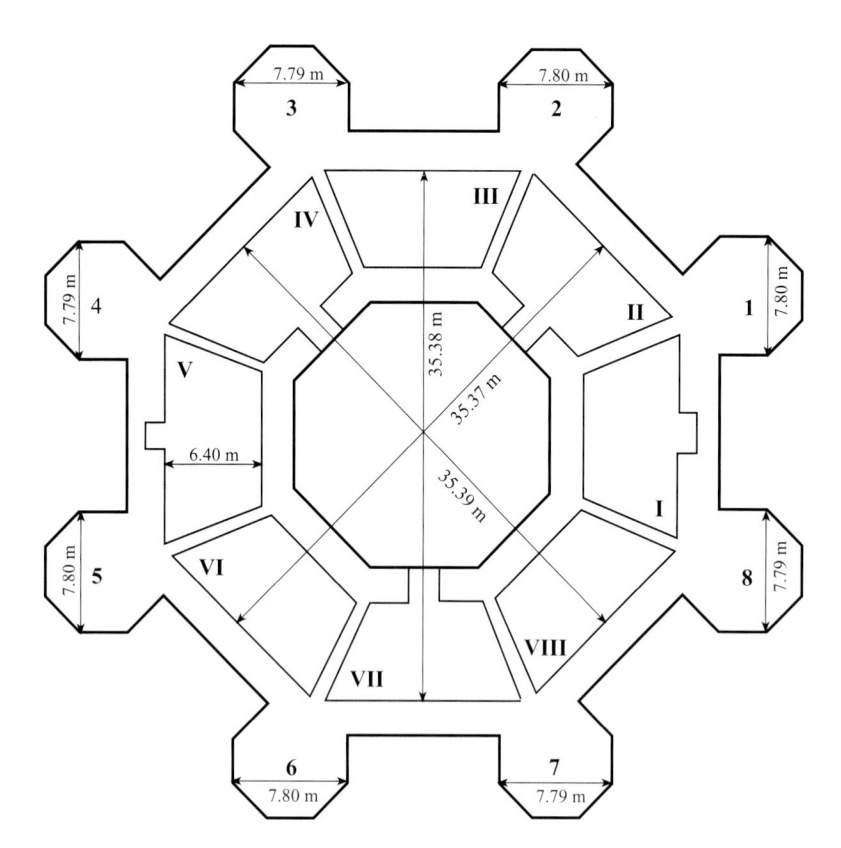

257 Castel del Monte. Main dimensions of the outer and inner walls of the citadel and the corner towers (survey data from the working group of W. Schirmer, Karlsruhe).

exactly 6.40 m, in two cases 6.39 m. The east wing (portal side) is an exception — although slight — here as well, being 6.42 m wide. Four sides of each tower are given as 3.23 m wide, one as 3.22 m, and the sides which meet those of the large octagon at right angles are 3.38 m (see also Fig. 244). The distances between towers are 10.35 m in three cases, 10.38 m in one case, and 10.31 m in one. There are two exceptions: the portal side, on which the distance between the towers is 20 cm greater (10.55 m), and the distance between towers 6 and 7 on the south side, which is 10.51 m. For the time being there is no explanation for these exceptions. The average values are as follows: 10.35 m if sides I (portal side) and VII are disregarded. If they are included, the average is 10.375 m. This is still an astonishing average for three towers; after all, in three instances the distance is exactly 10.35 m!

These are dimensional accuracies that seem unbelievable for a medieval European building. The exactness of the measurements is prerequisite to the unrestrained effect of the building, which presents itself symmetrically on all sides (Fig. 256). This precision — which has hardly been affected at all, even by the restoration work carried out on the citadel during past decades — certainly could

192   *Provenance of the Plan*

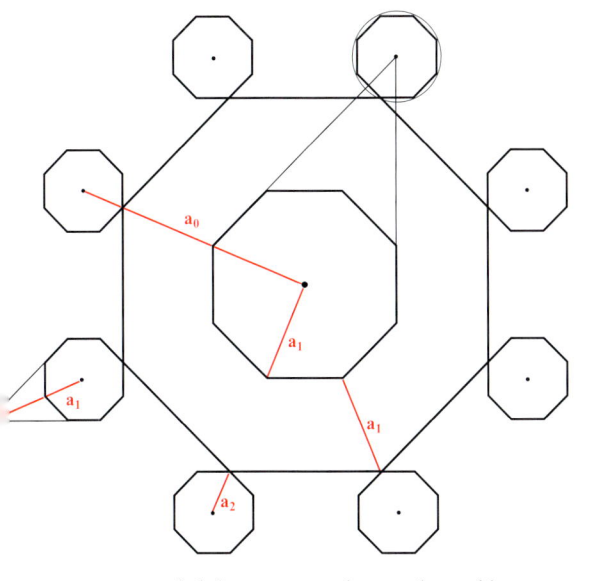

258 Castel del Monte. Relationship of basic dimensions with each other: diameter of the inner court to distance between inner court wall and inside of outer wall to width of the eight towers corresponds to $2:1:(\sqrt{2}-1)$. The ground plan of the citadel consists of three similar octagons.

$a_0 = \sqrt{2}+1$ Midpoint of the inner court to midpoint of a tower
$a_1 = 1$ Midpoint of the inner court to a corner of the inner court
$a_2 = \sqrt{2}-1$ Midpoint of a tower to one of the corners

not have been achieved without the use of precise technical methods of measurement. Consider alone the construction of the regular tower octagons with sides of precisely the same length.[260]

We discern an astounding regularity of the dimensional system within the scope of the three similar octagons, which, as we will see, are linked with one another as well by geometric rules.

The accuracy of the citadel's basic measurements with respect to its visible parts — outer walls, towers, inner court — is evidently independent, however, of the interior construction, including windows, doors, and stairways. This makes it clear that we are looking at *two* building units that are separate as far as the technical construction is concerned: The first is the *basic construction* consisting of outer walls, courtyard walls, and octagonal towers, with its extremely exact dimensions and correspondences, and the second is the *interior construction*, i.e., the arrangement of the interior rooms, their connections with each other and to the respectively finished tower interiors.

It is hardly conceivable that a design concept based on symmetry with extensive dimensional accuracy and probably strict geometric relationships should depart from it completely in the interior construction. The arbitrarily erected radial walls dividing the rooms and the resulting arrangement of the windows, which appear asymmetrical from the outside, are incompatible with that concept.

Let us consider first the *basic construction*. The attempt to find a method of construction for the two basic octagons, and above all a key to the size of the eight outer towers and their geometric connection to the central building, revealed (see Fig. 258) that the distance between opposite walls of the inner court is in a ratio of $2:1$ to the distance between the inner court walls and the inner surface of the outer wall of the large octagon (the measured dimensions are 17.8 m to 8.76 m = $2.03:1$!).[261] The width of the sides of the outer towers can be defined as $(\sqrt{2}-1)$.[262] This gives an overall proportion of $2:1:(\sqrt{2}-1)$.

The centers of the eight outer towers are found by extending the sides of the court octagon in both directions. In the outer intersections of these extensions lie the centers of the outer octagons, which are at the same time the eight points of a large star, the center of which is at

Provenance of the Plan 193

the center of the entire construction. This star octagon may be regarded as a basic geometric figure of Islamic ornamentation (see p. 129f.).

The size of the outer octagons is also defined by simple geometry. Each outer octagon is circumscribed by a circle whose center lies at one point of the star polygon and whose radius is equal to the distance of the inside corners of the large outer-wall octagon from the centers of the respective octagonal towers.

Now let us pursue further links between the three similar octagons. If we extend the eight sides of the eight corner towers in both directions, as done with the octagon of the inner court, the result again is eight-pointed star polygons. Surprisingly, one point of each polygon touches one point of the polygon on either side — proof of the exact geometry of the entire construction! Moreover, could there be any better confirmation of the strict connection between the three similar octagons (large octagon, inner court octagon, outer tower octagon)? The exact, geometrically definable ground-plan connections made it possible for Susanne Krömker to respresent the structure Castel del Monte graphically, based solely on the simple geometric formula of correlation (Fig. 253).

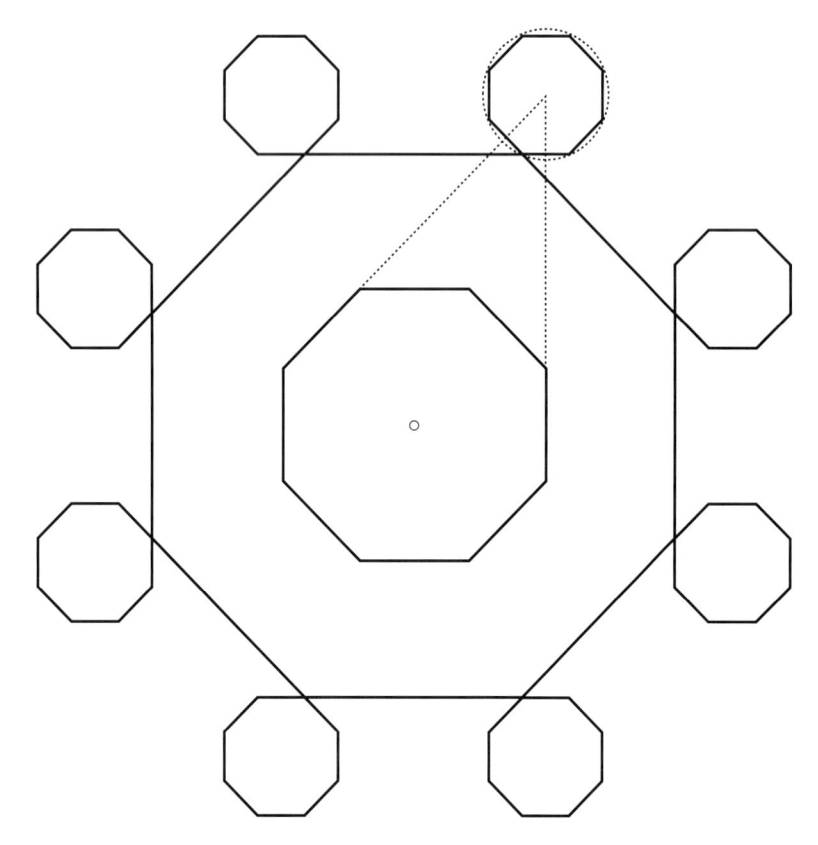

259 Castel del Monte:
The basic geometric concept.

194 *Provenance of the Plan*

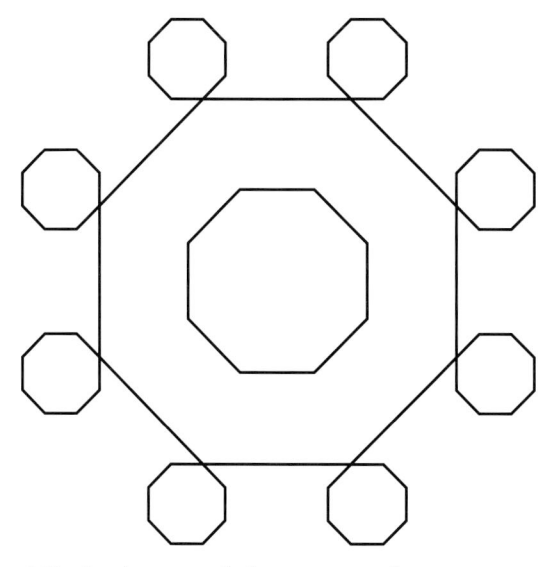

260 In the ground plan concept the towers are anchored geometrically, so to speak, as loops.

The diagram of the geometric connections between the two large octagons and the eight corner towers shown in Fig. 259 should be regarded as the *basic figure* for the plan of Castel del Monte. The corners of the line representing the inside of the large octagon's outer wall meet with the inside corners of the towers, owing to the fact that the contiguous sides of each tower are in alignment with the sides of the large octagon. This connection is all the more remarkable in that the inward-facing sides of the small octagons are not apparent, but disappear in the mass of the outer wall. They are executed only in the parts of the towers that project above the roof. Here they meet the inside corners of the large octagon,[263] which are also visible on the roof. With the outlines of the outer towers they form a type of loop (Fig. 260).

Let us summarize the geometric features of the basic concept of Castel del Monte and its plan that we have identified up to this point: 1. manifold symmetrical relationships, whose effect is intensified by 2. dimensional accuracies of the parts that are decisive for the form of the building: (a) outer wall, (b) courtyard wall, (c) corner towers, and 3. supplementation of the dimensional accuracies by fixed geometric relationships. These characteristic features do *not* apply to fortress construction in central Europe; they are, however, distinctive for Islamic architecture.

Symmetry is a distinguishing feature of Byzantine architecture, with its roots in Greek and Roman antiquity. From there, traditions passed into early Islamic architecture, noted for designs based on scientifically exact Euclidean geometry (cf. Sāmarrā, the mausoleum of Qubbat aṣ-Ṣulaibīya from 862 A.D., Figs. 170 and 171).

It should be borne in mind that Islamic mathematics, and above all geometry, had a long tradition, going back to the works of Euclid and Alexandrian geometricians such as Pappos.[264] This tradition had a lasting influence on Islamic architecture.

For some time now, we have been fortunate to have in our possession the publication of the Topkapi Scroll (Istanbul), which was commendably edited in 1996 by Gülru Necipoğlu. This work provides deep insights into Islamic geometry and its development. The Topkapi Scroll contains, in the words of the editor, "... far-reaching implications for the theory and practice of geometric

Provenance of the Plan 195

design in Islamic architecture and ornament. Created by master builders in the late medieval Iranian world, the scroll compiles a rich reportery [sic] of geometric drawings for wall surfaces and vaults. Not an isolated case, this important document belongs to a once-widespread Islamic tradition of scrolls in which geometric patterns ranging from *ground plans* [my italics] and vault projections to epigraphic panels and architectural ornament in diverse media appeared side by side."

The predominating geometric forms in this work are the circle, the square, and the octagon derived from it,[265] as well as the triangle and hexagon and compositions developed from them. The exact geometric drawings, in most cases centrally planned, form the basis for the arrangement of the muqarnas, which represent conveyance into the third dimension. The cleverly constructed geometric motifs testify to a particularly strongly developed two- and three-dimensional imaginative faculty.

Dr. Necipoğlu dates the Topkapi Scroll to the late fifteenth/early sixteenth century; its various parts are the result of a long development that was most impressive and most exemplary at its peak during the pre-Mongolian period.[266] In its scientifically geometric foundations and shapes it conformed with the strict observance of Sunnite religiousness.[267]

The shapes of the presented ornaments derive less from the inventiveness of craftsmen than from scientific geometric theory. In the hierarchy of that time in the Islamic area the mathematician/geometer occupied the highest position, succeeded by the practicing geometrician. The craftsmen followed these.[268]

This type of ornamentation experienced a lively development above all in the central Asian kingdom of Tamerlane, but also in the Maghreb, where even today designs for muqarnas in friezes and vaults are still executed in the traditional technique of the Middle Ages.[269]

Among the numerous illustrations contained in the Topkapi Scroll, two drawings in particular invite comparison with the plan of Castel del Monte: Fig. 97 (pp. 338 and 276) and Fig. 104 (pp. 342 and 279). Let us look at Fig. 104 (here Fig. 261): in the four corners of the basic square are four octagons, which form the centers of four four-pointed stars. These stars can logically be expanded to eight-pointed star polygons (Fig. 262). As the distance

261 Topkapi Scroll no. 104.

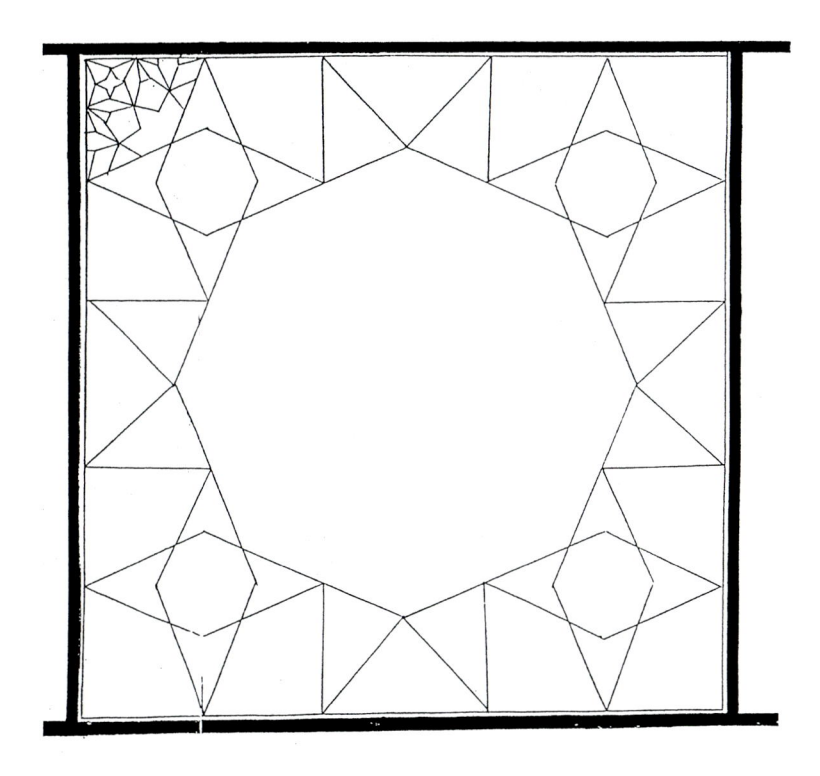

262 Topkapi Scroll no. 104. Expansion of the four-pointed to eight-pointed star polygons.

263 Topkapi Scroll no. 104. Schematic representation of rotation of the system by 45°.

between A and B is the same as that between B and C, an eight-part system can be produced from the four-part system by means of rotation by 45° (Fig. 263). It is distinguished by the fact that the points of each pair of neighboring eight-pointed stars touch, in the same manner as I elaborated with regard to the basic system of the eight towers of Castel del Monte (Fig. 254).

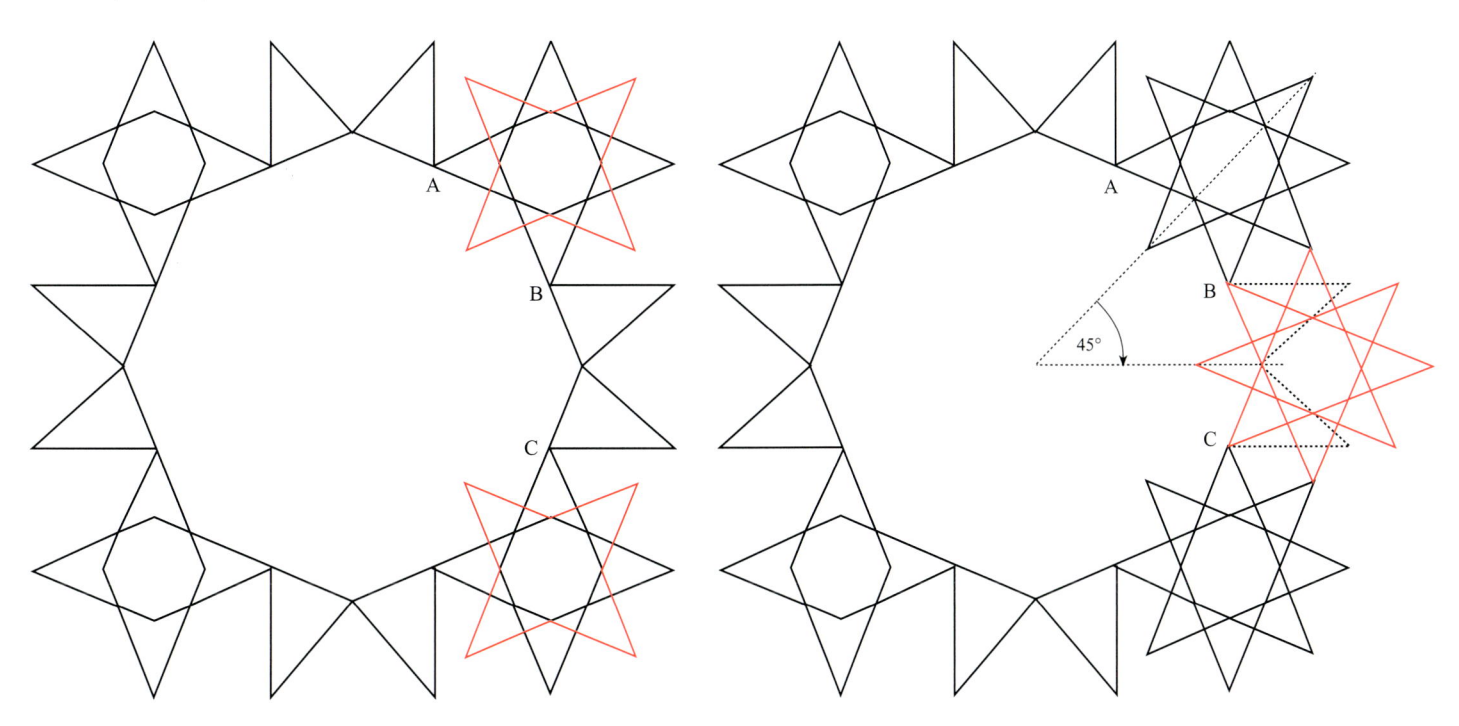

Provenance of the Plan 197

264 Topkapi Scroll no. 97.

If we now turn to the Topkapi Fig. 97 (here Fig. 264) and carry out the same rotation of the four-part system by 45°, we also obtain eight eight-pointed star polygons, with tangent point pairs.

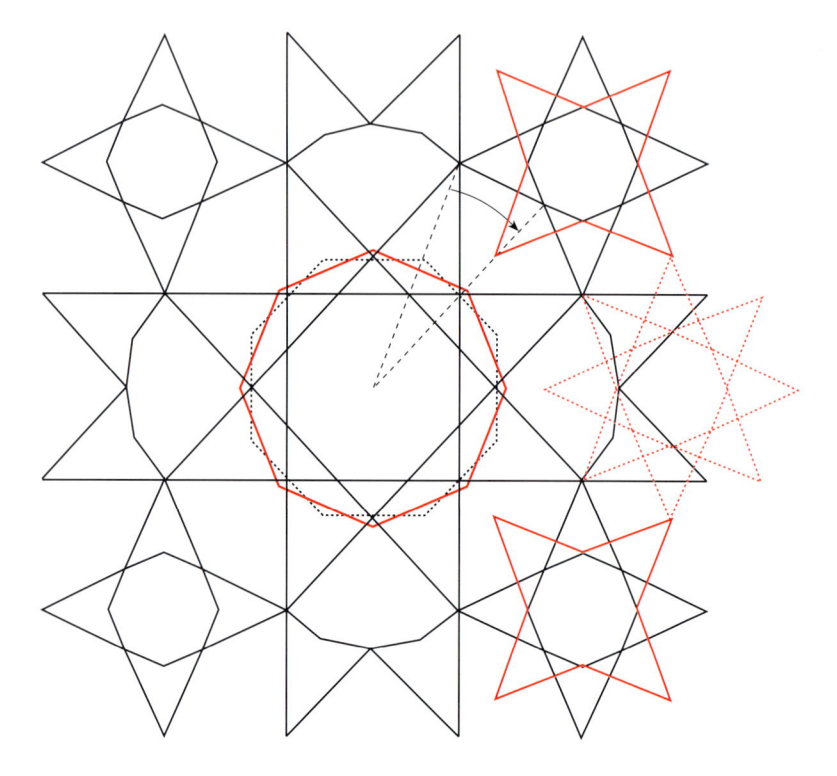

265 Topkapi Scroll no. 97. Rotation of the center octagon by 22.5°.

198 *Provenance of the Plan*

The Topkapi Fig. 97 shows in the center an octagon derived from two crossed squares; following a rotation by 22.5° (Fig. 265), its position in relation to the outer octagons surprisingly corresponds exactly to the relationship of the inner court octagon of Castel del Monte to the tower octagons. The same proportion holds as for Castel del Monte — $2 : 1 : (\sqrt{2}-1)$ (Fig. 266).

This dimensional *proportion* — not the actual dimensions — is one of the decisive features that connect the two plans.[270] Thus we find a confirmation of the principle elaborated by Ms. Necipoğlu of a geometric grid which represents the foundation for each individual geometric form.[271]

In the Maghreb the Islamic architectural techniques with their wealth of ornamentation have survived into the twentieth century.[272] In his book *Le Maroc et l'Artisanat Traditionelle Islamique dans l'Architecture*,[273] the French architect André Paccard described these techniques. He also drew attention to the 'secret of construction' kept by the Moroccans. It concerns the geometric grid system described above: "From one and the same grid, all design concepts can be made."[274]

Paccard reproduced a number of drawings from modern pattern books: "Even though the Moroccan pattern books betray some modern features, their traditional

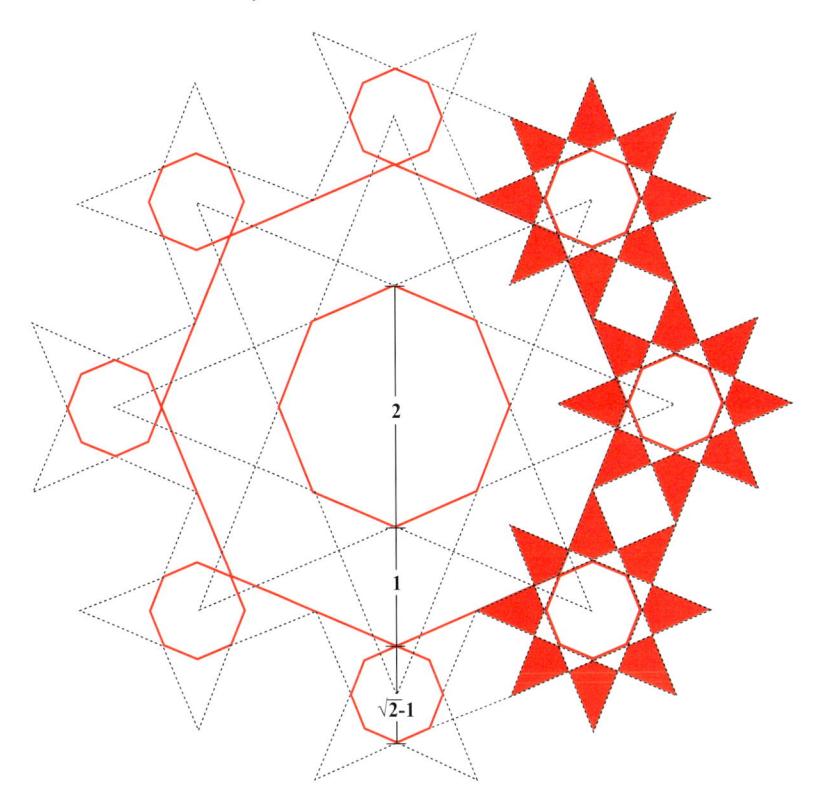

266 Topkapi Scroll no. 97.
The operations described in the text produce a picture identical to that in Fig. 254, also displaying the same basic proportions.

Provenance of the Plan 199

two- and three-dimensional geometric designs once again testify to the role of drawings in assuring the perpetuation of a mode of architectural decoration that dates back *at least* to the fourteenth or fifteenth century, *if not earlier.*"[275] One of the drawings there is of particular interest in connection with the plan of Castel del Monte: it is reproduced here as Fig. 267. It is identical to the drawing in Fig. 268 and corresponds in all basic proportions to the plan of Castel del Monte!

In her studies on al-Kashi's methods of measuring the muqarnas, Yvonne Dold-Samplonius examined modern muqarna constructions in Morocco and obtained projection drawings from the architects there, which are related in principle to those of the Topkapi Scroll[276] (here Fig. 269). It is interesting that a system of two similar octagons set into one another appears here as well. The larger (outer) octagon shows eight small octagons that are, however, directed inward.

After all this, there can be no doubt that the plan of Castel del Monte is derived from that geometric world that we have been able to identify. Above all, the identity of the plan (Fig. 254) with the geometric systems of the Topkapi drawings 104 and 97 (here Figs. 261 and 264) is compelling.

The evidence presented here of the connection of Castel del Monte to, or rather the origin of its plan from, the tradition of Islamic construction drawings stems from

267 El Bouri: Patterns for octagonal and star-shaped muqarna vaults, Morocco, twentieth century (from Paccard 1979, 1:296, Fig. 3. Topkapi Scroll, p. 25, Fig. 50).

268 Copy made from Fig. 267 with inner, outer, and tower octagons drawn in red and standing in the proportion $2 : 1 : (\sqrt{2} - 1)$, exactly as in Castel del Monte!

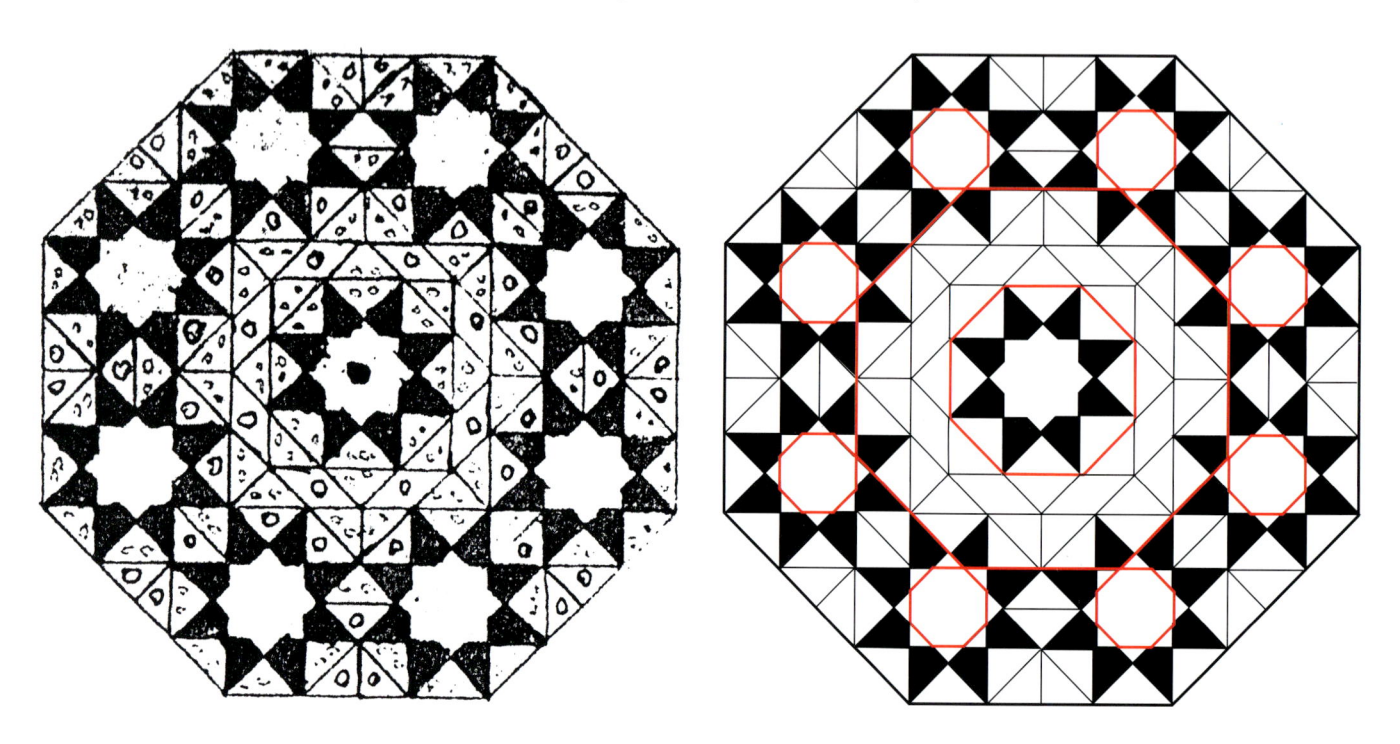

Provenance of the Plan

269 A Moroccan plan for the construction of muqarnas (after Yvonne Dold-Samplonius, 1996, p. 71, Fig. 12). We see the inner and outer octagons in the same proportion as shown in Fig. 226. The small octagons, however, are not attached to the large octagon but rather 'folded' inward. They are thus in a different proportion to the larger octagons.

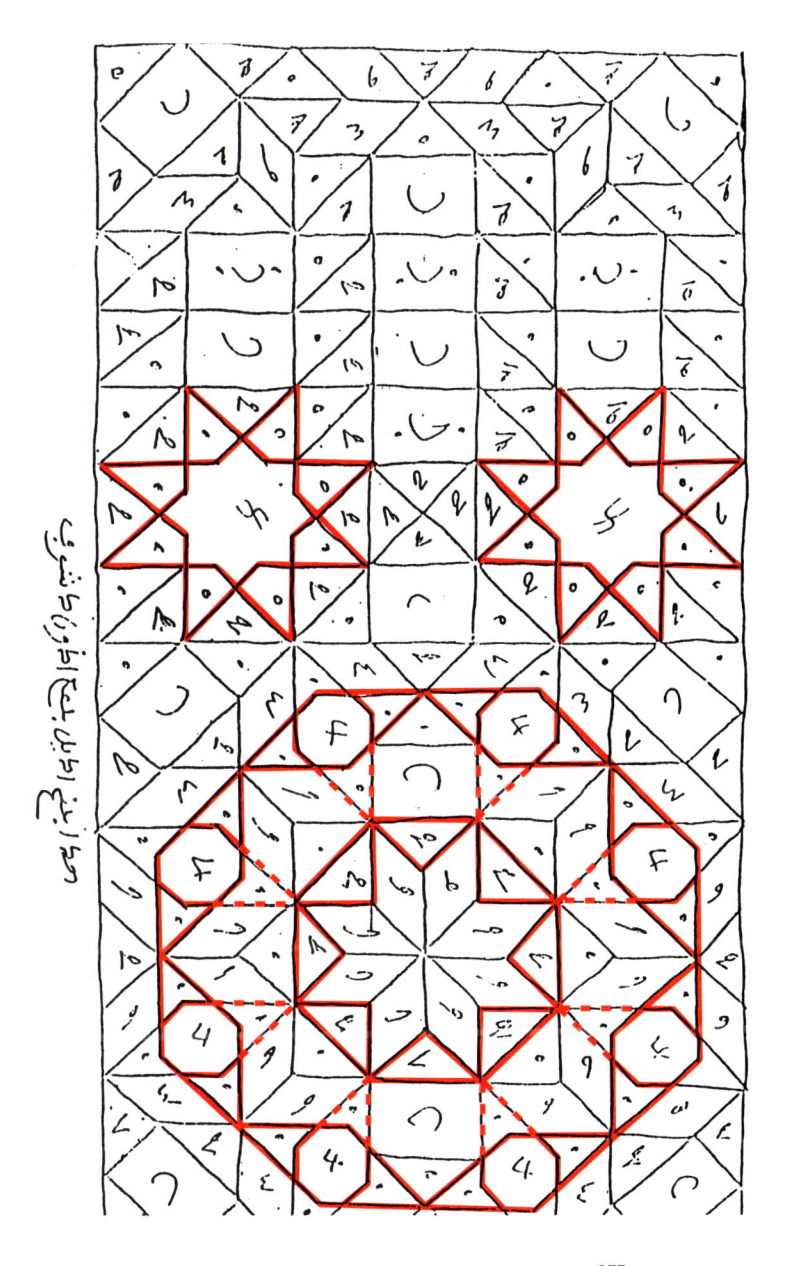

a consideration of the building in its entirety.[277] The course of the citadel's outer wall and the width of the towers — impressively exact to the centimeter — represent the foundations for this consideration of measuring techniques. As important as the constructional details pertaining to the interior certainly are (see p. 193), inasmuch as they are evidence here, too, of connections with the Oriental-Islamic cultural complex, proof that the plan for the basic construction had its source in the Islamic tradition of geometric form and architectural practice goes beyond these as a decisively important fact on a higher level. It definitively separates Castel del Monte from central and northern European architectural tradition.

Provenance of the Plan 201

As suggested at the beginning, the — to a great extent visible — characteristics of symmetry of the building dimensions, the resulting strict parallelism of geometrically corresponding lines and wall surfaces, and the two-dimensionality of design all speak for such an Islamic architectural principle.

In view of Frederick II's great liking for the Arab sciences and particularly mathematics and astronomy, this discovery is not surprising. It is interesting in this respect that Frederick's contacts belonged to the Arab region of the Sunnite Ayyubides (see above), who had succeeded the Shiitic Fatimids. Frederick's great interest in the art of that world has recently been reconfirmed by the interpretation of a document which reports on the visit of one of his emissaries to the pyramids of Giza.[278]

We are not aware of any one architect who was responsible for Castel del Monte.[279] There are various conceivable ways in which Muslim architects may have influenced the design of Castel del Monte directly or indirectly. Several characteristic features of Muslim architecture in the interior (see note 278) speak for the direct participation of Muslim building experts.

It must not be forgotten that a great deal of interplay took place in the medieval border areas between the regions of Islam and central Europe, both in Spain and in southern Italy. This was much stronger in Spain, where the Arabs remained for 800 years, and is clearly visible in many areas even today, not only in Andalusia itself. In Sicily, where Moslem rule lasted just under 250 years, its influence was less persistent; presumably, however, there were still Arab architects in Palermo during Frederick's lifetime. On the other hand, perhaps Frederick's contacts in the territory of the Ayyubides provided ideas and architectural assistance.

In any case, the manifest connection between the plan of Castel del Monte and Islamic decorative art, which is so closely tied to the architecture, is final proof of the origin of the design for this unique work of medieval architecture in Europe. The stonemasons' lodge that carried out the construction subjected itself to the strict nature of the design, especially in its outer appearance. In the interior the blind arcades, the form of the capitals, and numerous other details show the hand of the Cistercians and thus central European characteristics.[280]

Let it be said that two separate design principles can be recognized in the overall architectonic appearance of the building: the (Islamic) *basic construction* and the (Cistercian?) *interior*. This systematic division of the overall design of Castel del Monte into (a) an 'Arab-Islamic' plan with astonishing dimensional accuracy[281] and (b) an interior that conforms to Cistercian, i.e., central European, architectural tradition clarifies much that has up to now been attributed to the building's enigmatic nature.

Ad (a): The impressive symmetry of the design concept cannot be explained in terms of central European architectural tradition, but rather in terms of the traditions of classical, Byzantine, and Islamic architecture, in which symmetry was the central principle of order.

Ad (b): The interior construction with the inexactly aligned internal walls of the trapezoid-shaped rooms conforms in no way to exact geometric convention, but rather to the practice of central European stonemasons' lodges, which managed without strict geometry (see the lodge book of Villard de Honnecourt).

We will not be going wrong if we assume Frederick II's personal participation in the design of this building, which was intended to symbolize his conception of the nature of his rule — a participation which most probably exceeds the contribution he is known to have made to the bridge gate in Capua (see p. 68, note 69). H. Martin Schaller[282] has discovered itineraries of Frederick II for that period of time which either touched on Castel del Monte or passed by it very closely. It is logical to interpret these visits as serving to check on the progress being made on construction.

Let us return to the starting point of our considerations about the manifoldly symmetrical structure of Castel del Monte, a structure unmatched in the history of architecture. The appearance of this geometric figure is enhanced by the purity of its realization, which does without any kind of additional decorative element. We found the basic

Provenance of the Plan 203

idea for the geometric concept, with its 'rotating' squares and the firm connections to the centers of the corner towers that result from extending the sides of the inner octagon, in a splendidly constructed symmetrical ornament in a prominent position in the Alhambra, as well as in the compass rose of the *Carta Pisana*.

Thus the geometric experience of early medieval Islamic mathematics, fed by Hellenism, fuses with classical-European architectural tradition, which turned the idea of the symmetrical ornament group into an overwhelming building. The Hohenstaufen architecture of southern Italy which precedes Castel del Monte already displays this geometric-stereometric structural constituent clearly. Throughout the centuries and into our own time, hardly a building can be found that realizes a geometric concept so absolutely as Castel del Monte.

We know from various examples that symmetry is not shaped by man alone, but is also a characteristic of nature, both animate and inanimate. In its different forms

270 Castel del Monte, Apulia: View from the southeast.

204   *Provenance of the Plan*

and variations in nature, as in art, symmetry is a structural element that can be defined mathematically. The relationship of ordered, structured nature with art is not surprising, because art has always been at once the imitation and the exaggeration of nature. With the entire spectrum of its forms, symmetry has repeatedly been employed as a means of style in art — this is particularly evident in architecture. Did the artist then discover the aesthetic quality of symmetry in nature, or is there even a common source of symmetry's aesthetic quality in nature and art?

What enables man to intuitively grasp the basic mathematical structure of the world and to reconstruct and reexperience it seems to me aptly expressed in the lines from Plotin, translated by Goethe:

> If the eye were not like the sun
> The sun it could not see;
> If God's own power were not within us
> How could the divine enrapture us?[283]

There appears to be a process of recognition, an *anagnórisis*, corresponding to Plato's *anámnesis*, the recollection of the world of ideas, to which he also assigns mathematics. Why else is an exactly defined proportion such as the golden section or the five platonic bodies — called platonic because Plato was the first to define them — spontaneously perceived as being beautiful — even by people who are unaware of the underlying mathematical principles? Castel del Monte thus appears to us as a living witness to the mathematical nature of aesthetics.

VII

Interpretation

The unusualness of its exterior form and the splendor of its interior appointments, unequaled by any building of that time, are an important indication of the status and significance that the emperor intended this citadel to have. This speaks against it being a structure for 'normal' purposes, such as a fortress or a simple hunting lodge. The chosen building site lies on a cone-shaped hill, which made the castle-like citadel visible from all sides, even when the woods were denser, standing as it did on a natural socle.

The form of the manifoldly symmetrical octagon of Castel del Monte takes up the symbolic power of the Carolingian octagon and the octagonal imperial crown.[284] It reflects the ground plan of the Church of the Nativity at Bethlehem (Fig. 271) as well as of the Dome of the Rock (p. 137, Fig. 205). We know how deeply impressed Frederick was by the experience of being crowned as emperor in the Pfalzkapelle of Charlemagne (Figs. 272 and 275).[285] The inner connection may have been strongly felt in the presence of the octagonal chandelier with eight portals, eight towers, and 48 lights (Fig. 274), donated to the Pfalzkapelle in 1165 by his grandfather Frederick Barbarossa.[286]

We recognize the octagon as the leitmotif of the imperial symbology that surrounded Frederick II. As we have seen, in the design of Castel de Monte the octagon motif experiences manifold extension and elevation in the form of geometric entwinements, whereby the aesthetic components of mathematics become clearly visible. We do not know whether Frederick II ever visited the citadel. Hans Martin Schaller argues plausibly that the emperor's scouting trips through the Murge area would have passed by the citadel during its construction (see above, p. 203). Furthermore, a Hohenstaufen statute exists from the year 1246, according to which the residents of Monopoli, Bit-

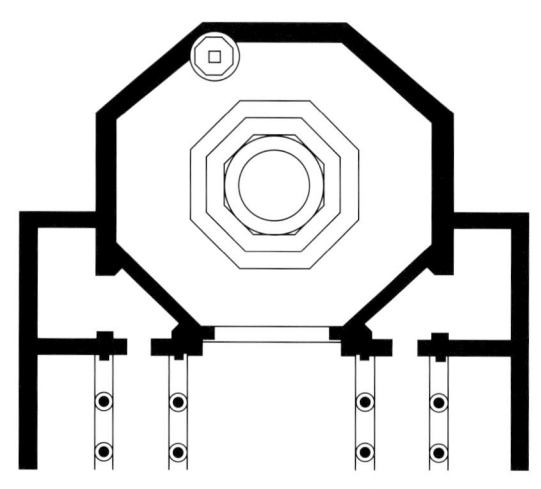

271 Bethlehem: Ground plan of the Church of the Nativity (from the time of Constantine the Great, 326 A.D.; simplified from: Gerhard Kroll, *Auf den Spuren Jesu*, p. 54, Fig. 30, Stuttgart 1983).

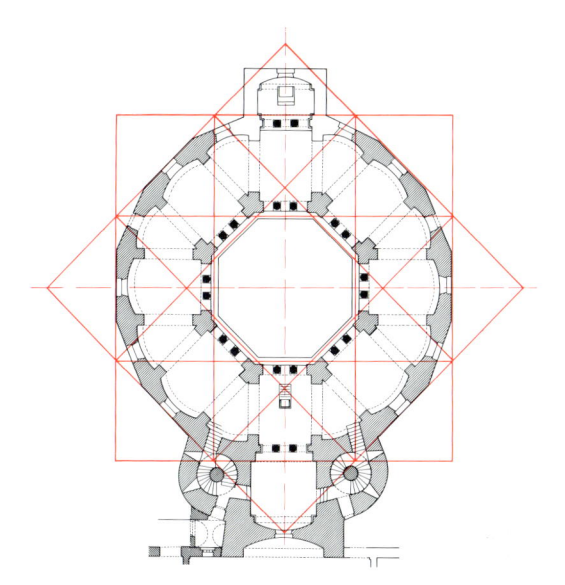

272 The 'Octagon' in Aachen: Plan of the upper story of the Carolingian Pfalzkapelle (ca. 800) with eight-pointed star drawn in.

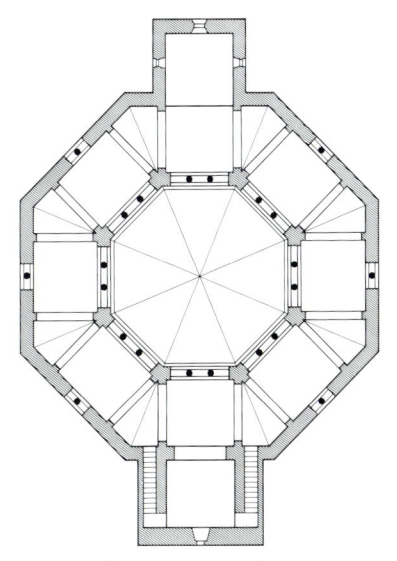

273 Ottmarsheim, Haut-Rhin: Former abbey church, later collegiate church (1030). Plan of the upper story (see also: Jakob Burckhardt, 'Die Kirche zu Ottmarsheim im Elsaß' (1844), in: *Jakob Burckhardt, Gesamtausgabe*, vol. 1: *Frühe Schriften*, Basel 1930, p. 297 ff. with illustration).

teto, and Bitonto were obligated to carry out repair work on the citadel.[287] From this it must be concluded that the citadel had been completed to a certain degree by this time and was also most likely partially inhabitable.

The outward appearance of the citadel is determined by its symmetrical relationships; the lustrous building with the Islamic-geometric ground plan based on the octagon in threefold variation has the portal directed along the main axis toward the east. This and the absence of a chapel speak clearly for a secular purpose. The north-south axis, running at a right angle to the main axis, meets the spire of the cathedral of Andria in the north, in whose crypt both of the emperor's wives, Isabella of England and Isabella of Brienne, were laid to rest. Andria was the town that had always been loyal to the emperor — Andria Fidelis.

On the basis of his studies on classical systems of measurement, Florian Huber[288] was recently able to offer serious support for an interpretation that had already been voiced by C. Meckseper, W. Krönig, and H. M. Schaller[289]: Castel del Monte as the symbol of divine Jerusalem. Huber equates the side of the square that surrounds Castel del Monte with 100 ancient ells or Hebrew royal ells, the latter being equal to 51.83 cm. The New Temple in Jerusalem also measures 100 royal ells.[290] This implies a symbolic equation of Castel del Monte with the New Temple of Jerusalem, the city of David. This symbolic connection has a far-reaching significance. Frederick II was the only Holy Roman emperor who, on March 18, 1229, had placed the crown of Jerusalem on his own head and thus considered himself a descendent of David, who — like Christ — was born in Bethlehem. Reference to King David was of great importance in the Middle Ages.

We are reminded of the wording of the letter written by Frederick II in August 1239 to Iesi, the town of his birth: "Following the promptings of Nature, we are compelled and obliged to embrace with deepest love Iesi, noble city of the Marches, illustrious very beginning of our beginning, where our cradle brightly shone, in order that it not disappear from our memory. May its place, and our Bethlehem, the land and origin of Caesar, remain deeply rooted in our breast."

With the interpretation of Castel del Monte as a symbol of heavenly Jerusalem and the consequent associa-

Interpretation 207

274 The 'Octagon' in Aachen: Barbarossa chandelier (1165–1170) in the Carolingian Pfalzkapelle.

tions to the emperor himself as a descendent of King David, the citadel attains a significance corresponding to the emperor's discernible conception of his divine mandate to rule following his return from the Holy Land.[291]

The connections to Andria mentioned above expand the local network of relationships, as does the proximity to his residence in Foggia, which — judging from the remains — was splendidly appointed. Also belonging to Frederick's 'area of residence' in northern Apulia, finally, is the fortified citadel of Lucera with the settlement of his Saracen troops.

If Palermo was the place of residence in his southern Sicilian kingdom, the residential area of Foggia, with the symbolically exalted, dazzling Castel del Monte, could well have been his Bethlehem, center of the emperor's concept of himself as the divinely appointed earthly ruler over the Christian world. The location of this 'residence' on the southern border of the Papal States underlines this symbolic character.

G. Wolf has theorized that, following the quarrel with his son, King Henry VII, Frederick II may have considered keeping the imperial crown jewels *(Reichsinsignien)* at Castel del Monte, and thus near him.[292] This is, for the present, his hypothesis, but it fits the importance of Castel del Monte as a residence and center of Frederick II's sovereign power.

The last ten years of the emperor's life, during which the building was constructed, were burdened with political setbacks and human disappointments, and the subsequent end of his dynasty put a sudden stop to the imperial

275 The 'Octagon' in Aachen: Inside view of the Carolingian Pfalzkapelle (ca. 800) with Barbarossa chandelier.

208   *Interpretation*

276 Aerial photograph of Castel del Monte.

ideas and plans. We are left with an architectural masterpiece of unusual ingenuity with regard to its combination of intellectual concepts of varying provenance, put into effect with geometric strictness and artistic inventiveness. With Castel del Monte, Hohenstaufen architecture was carried to a glorious zenith after an astonishingly short period of development. It had grown out of the traditions of Magna Graecia and the Roman Empire, the Greek-Hellenistic heritage of the Islamic sciences, the spirit of Norman-Arab Sicily, and central European Gothic — a truly universal conception.

Transcription Table

The transcription of Arabic and Persian words and names follows the rules of the German Oriental Society (Deutsche Morgenländische Gesellschaft). Long vowels are indicated by macrons (ā, ī, ū).

č	correspondends to ch in *Dutch*
ḍ	stressed, explosive (glottal) *d*
ġ	uvular *r*
ǧ	*j* as in *jungle*
ḥ	voiceless, guttural *h*
ḫ	hard *h*, corresponding to Scottish *loch* or German *Bach*
q	*k* sound, formed at back of the throat
Š, Ş	*sh* as in *shine* (š is used only in Turkish words or names)
ṣ	stressed, explosive (glottal) *s*
ṭ	stressed, explosive (glottal) *t*
ṯ	voiceless *th* as in *thing*
y	as in *yet*
z	voiced *s*, corresponding to *z* in *zebra*
ẓ	stressed, explosive (glottal) *z*, or voiced glottal *t*
ʼ	glottal stop (as in German *be'enden*); also may indicate a dropped letter
ʻ	voiced, guttural *a*-sound made by compressing the throat

Notes

Literature cited in abbreviated form is listed in detail under References beginning on p. 225.
A superscript number after a year indicates which edition is cited.

1 Wolfgang Krönig, 'Staufische Baukunst in Unteritalien', in: *Beiträge zur Kunst des Mittelalters* (contributions presented at the First German Congress of Art Historians at Schloß Brühl, 1948), Berlin 1950, p. 33 ff.

2 Kurt Vogel, 'Fibonacci, Leonardo, or Leonardo of Pisa', in: *Dictionary of Scientific Biography*, edited by Charles Coulston Gillispie, New York 1971, vol. IV, pp. 604–613. See also: B. L. van der Waerden, *A History of Algebra*, Berlin/Heidelberg/Tokyo 1985, pp. 32–46; finally, the excellent paper must be mentioned that Rashed Roshdi read at the 'International Seminar on Frederic II' in Erice (Trapani, Sicily) in September 1989: 'Fibonacci et les Mathématiques Arabes', in: *Actes du Colloque sur Frédéric II. et les Savoirs*, Palermo 1990.

3 The development of the Arab influence on the West had begun earlier, earlier even than the generally accepted interpretations of C. H. Haskins have suggested to date — for instance, in his *Studies in the History of Mediaeval Science*, Harvard 1927. See also: Daffā', 'Alī 'Abd Allāh and John J. Stroyls, *Studies in the Exact Sciences in Medieval Islam*, New York 1984.

4 The work by Gino Chierici, 'Castel del Monte', as part of the edition *I Monumenti Italiani, rilievi, raccolti a cura della Reale Accademia d'Italia*, first fascicle, Rome 1934, unfortunately lacks the reliable measurements that one would expect to find in a building survey.

5 J. W. von Goethe, *Westöstlicher Diwan, Buch der Sprüche*, in: *Goethes Sämtliche Werke*, Insel edition, vol. II, Leipzig 1926, p. 693.

6 *Ryccardi de Sancto Germano Notarii Cronica*, Muratori, Rer. Ital. Script. Nova Ser. vol. VII, Paris II (edited by C. A. Garufi), 1938, Script. 32, 349.

7 Peter Cornelius Claussen, 'Die Statue Friedrichs II. vom Brückentor in Capua (1234–1239)', in: *Festschrift für Hartmut Biermann*, Weinheim 1990, pp. 19–39, Figs. 1–25 on pp. 291–304. See also Carl Arnold Willemsen, *Kaiser Friedrichs II. Triumphtor zu Capua. Ein Denkmal hohenstaufischer Kunst in Süditalien*, Wiesbaden 1953, p. 37 f., plate 37.

8 See M. L. Finley et al., *A History of Sicily*, London 1986, p. 51 f.

9 Francesco Gabrieli, 'Friedrich II. und die Kultur des Islam', in: *Stupor Mundi*, Darmstadt 1982[2], pp. 129 and 271.

10 The fortunate discovery of a pre-Norman, Arab citadel in Mazzallaccar near Sambuca in western Sicily by Anna Maria Schmidt confirms this. Schmidt's research results are recorded in: 'La Fortezza di Mazzallaccar', in: *Bolletino d'Arte*, series V, vol. 57, 1972, pp. 90–93. See also here, p. 53 ff.

11 Salvatore de Renzi, *Storia documentata della scuola medica di Salerno*, revised edition Milan 1967, p. 196 f.

12 Gabrieli 1982[2], p. 288.

13 Liber decretalium extra Decretum Gratiani.

14 Gunther Wolf, 'Kaiser Friedrich II. und das Recht', in: *Zeitschrift der Savigny Stiftung für Rechtsgeschichte*, RA 102 (1985), p. 327 ff., particularly p. 338 ff.

15 Francesco Torraca, 'Le origini. L'età sveva', in: *Storia della Università di Napoli*, edited by Gennaro Maria Monti, Naples 1924. In his 'Chronica priora', the secretary Riccardo da San Germano handed down the text of the founding letter from Syracuse, dated June 5, 1224.

16 Emperor Frederick II, *Über die Kunst, mit Vögeln zu jagen*, transcribed and edited, in cooperation with Dagmar Odenthal, by Carl Arnold Willemsen, 2 volumes, Frankfurt a. M. 1964; annotation volume by Carl A. Willemsen, 1970.

17 Regarding Leonardo of Pisa, also called Leonardo Fibonacci, see note 2.

18 Francesco Torraca, *Studi su la lirica italiana del duecento*, Bologna 1902, p. 238.

19 Taken from a letter sent from St. Avold on August 21, 1215, to the general chapter of Cistercian abbots. See Klaus J. Heinisch (editor and translator), *Kaiser Friedrich II. in Briefen und Berichten seiner Zeit*, Darmstadt 1978[6], p. 33 ff.

20 Ibid., p. 81 ff.
21 Eberhard Horst, *Friedrich der Staufer*, Düsseldorf 1975, p. 12.
22 Quoted loosely from Hans M. Schaller, *Kaiser Friedrich II., Verwandler der Welt*, Göttingen 1971², p. 87 ff.
23 Renate Wagner-Rieger, *Die italienische Baukunst zu Beginn der Gotik*, part II: 'Süd- und Mittelitalien', Graz/Cologne 1957, p. 169.
24 Carl Arnold Willemsen, 'Die Bauten der Staufer', in: cat. *Die Zeit der Staufer*, vol. 3, Stuttgart 1977, p. 151, Fig. 6.
25 Wagner-Rieger 1957, p. 166.
26 Jean-Louis-Alphonse Huillard-Bréholles, *Historia Diplomatica Friderici Secundi sive Constitutiones...*, vol. 1, part 1, Paris 1852–1861, p. 509.
27 *Architettura sveva nell'Italia meridionale, Repertorio dei castelli federiciani*, Palazzo Pretorio, Prato, March–September 1975 (edited by Arnaldo Bruschi and Gaetano Miarelli Mariani), Florence 1975 (hereafter cited as *Repertorio*), p. 186 ff.; Walter Hotz, *Pfalzen und Burgen der Stauferzeit. Geschichte und Gestalt*, Darmstadt 1981, p. 312 f., Fig. Z 180, plates 195–197.
28 Giuseppe Agnello, *L'Architettura sveva in Sicilia*, Collezione Meridionale, Rome 1935, p. 74 ff., Figs. 42, 43.
29 Ibid., pp. 50 and 55, Figs. 19, 20, and 23. Regarding the bronze ram: J. W. von Goethe, *Italienische Reise* (Zweiter Teil), Palermo, Mittwoch den 11. April 1787, in: *Goethes Sämtliche Werke*, Insel edition, vol. 4, Leipzig 1926, p. 262. Goethe saw both rams; the second was lost in the revolutionary disturbances of 1848. — Pirro Marconi, 'Note sull'Ariete del Museo di Palermo', in: *Bollettino d'Arte del Ministero dell' E. N.*, September 1930, pp. 132–143. — Agnello 1835, p. 56, Fig. 24.
30 P. Paolini, 'Nuovi aspetti sul castel Maniace di Siracusa', in: *Atti del III congresso di architettura fortificata (1981)*, Rome 1985, pp. 215–222. See also: W. Krönig, 'Prefazione' to: Agnello 1935 (reprint, Syracuse 1986).
31 See note 104.
32 Agnello 1935, p. 86, Fig. 64.
33 Ibid., p. 63, Fig. 31.
34 Willemsen 1953. — Friedrich Wilhelm Deichmann, 'Die Bärtigen mit dem Lorbeerkranz vom Brückentor Friedrichs II. zu Capua', in: *Von Angesicht zu Angesicht, Michael Stettler zum 70. Geburtstag*, Bern 1983, pp. 71 ff. and 75.
35 *Repertorio*, p. 174 ff.; Hotz 1981, p. 309 f., Fig. Z.
36 Schmidt 1972, p. 92, second column; cf. also C. Diehl, *Manuel d'art byzantin*, Paris 1925/26, p. 184 and idem, *L'Afrique byzantine I*, Paris 1896, p. 145.
37 Cf. also Carl C. Willemsen, 'Die Bauten der Hohenstaufen in Süditalien', in: *Arbeitsgemeinschaft für Forschung des Landes Nordrhein-Westfalen*, 25th issue: Geisteswissenschaften, Cologne/Opladen 1968, p. 42, Fig. 25.
38 *Repertorio*, p. 170 ff.; Hotz 1981, p. 310 f., Fig. Z 179, plates 93/94.
39 Concerning the ground plan, cf. Krönig 1950, p. 393, Fig. 100.
40 Willemsen 1977, p. 152.
41 Agnello 1935, p. 443, Fig. 292.
42 Krönig 1950, p. 28 ff. Cf. also: Stefano Bottari, *Monumenti svevi di Sicilia*, Palermo 1950, and recently: Antonio Cadei, 'Architettura federiciana', in: *Nel Segno di Federico II, Atti del IV Convegno Internationale di Studi della Fondazione Napoli Novantanove*, Naples 1989, p. 143 ff.
43 Willemsen 1968, p. 44 f.
44 Ibid., p. 45, Fig. 28. — Cord Meckseper, 'Castel del Monte. Seine Voraussetzungen in der nordwesteuropäischen Baukunst', in: *Zeitschrift für Kunstgeschichte* XXXIII, 1970, pp. 211–231, returns once more to the idea that the origin of Gothic in Sicily was ultimately French — particularly in regard to Castel del Monte. This origin must necessarily be expressed not only in the details and in the Cistercian structural forms but also in the layout as a whole. He then, on the basis of his knowledge of military architecture, cites examples on p. 213 ff. of his contribution that are reproduced again by W. Krönig in: Émile Bertaux, *L'Art dans L'Italie Méridionale*, an update of the work by Émile Bertaux under the direction of Adriano Prandi, vol. 5, Rome 1978, e. g., p. 946, Fig. 105. However, they are so fundamentally different in their geometric structure from the Hohenstaufen architecture of Lower Italy — consider for instance Boulogne-sur-Mer (ibid., Fig. 7) — that no connections are visible here.
45 Cf. Finley 1986, p. 60.
46 Mohamed Talbi, *L'Emirat aghlabide*, Paris 1966, p. 416. Courtesy of H. Halm.
47 The geographer Ibn Ḥauqal came from Nisibis (Arabic Nuṣaibīn) in northern Mesopotamia, on the present-day border between Turkey and Syria. The dates of his birth and death are unknown. His journeys began in 943 A.D., taking him through North Africa, Spain, the Sahara, Egypt, Armenia, Iran, and Central Asia. His 'The Shape of the Earth' appeared in several editions, the final one probably about 988 A.D. A French translation was published by J. H. Kramers and G. Wiet, *Configuration de la terre*, Beirut/Paris 1964 (information from H. Halm).
48 From the translation by Kramers/Wiet 1964, p. 120.
49 On this point see Willemsen 1968, p. 42 f.
50 Cf. Henri Stern, 'Notes sur l'architecture des châteaux omeyyades', in: *Ars Islamica* (The Department of

Fine Arts), vols. XI–XII, University of Michigan Press, Ann Arbor 1946.

51 Cf. Schmidt 1972, p. 90ff., and Willemsen 1968, p. 42, Fig. 25, as well as Krönig, in: Bertaux 1978, p. 944, Fig. 104 and idem in: Guido di Stefano, *Monumenti della Sicilia Normanna*, second edition revised and enlarged under the supervision of W. Krönig, Palermo 1979, p. 136, plate CC. Further in general G. Agnello, 'Problemi ed Aspetti del Architettura Sveva', in: *Palladio*, X, 1960, pp. 37–47. Finally: S. Bottari, 'Ancora su le Origini dei Castelli Svevi della Sicilia', in: *Atti del Convegno Internazionale di Studi Fredericiani in Palermo 1950*, Palermo 1952, pp. 501–505.

52 Dieter Mertens, 'Castellum oder Ribāṭ? Das Küstenfort in Selinunt', in: *Mitteilungen des Deutschen Archäologischen Institutes Istanbul*, vol. 39, 1989, pp. 391–398, plate 37.

53 Willemsen 1968, p. 42, Fig. 25. Cf. also Mertens 1989, p. 396f.

54 Finley 1986, p. 48f.

55 Cf. the beautiful photograph of al-Ḥarāna from: Henri Stierlin, *Architektur des Islam vom Atlantik zum Ganges*, Zurich/Freiburg 1979, pp. 75, 76, Fig. 35 (here p. 53, Fig. 59).

56 This concerns not just any arbitrary numbers, but only those groups of numbers which can be rendered as the Pythagorean equation with natural whole numbers (see John H. Conway and Richard K. Guy: *The Book of Numbers*, New York 1996, p. 172, Figs. 6–8). Pythagorean triangles and Pythagorean triples are of great importance, not only for theoretical mathematics but also for the practical tasks of surveying and architecture. Their rules were known long before Pythagoras in Babylonia, Egypt, India, and China. B. L. van der Waerden devoted no fewer than 35 pages to them in his book *Geometry and Algebra in Ancient Civilisation*, from 1983. Regarding Leonardo Fibonacci, see Vogel 1971, pp. 604–613, in particular p. 608, right-hand column. Cf. here also note 2.

57 Cf. Nils G. Wollin, *Desprez en Italie*, Malmö 1935. Louis Jean Desprez is called simply a 'draftsman' in the more recent literature. However, he was — and this increases the authentic value of his architectural drawings considerably — an architect and graduate of the Académie Royale de l'Architecture in Paris, from which he received the Grand Prix de l'année on August 26, 1767. He traveled to Rome, Lower Italy, and Sicily and left numerous documentarily valuable drawings, the reliability and accuracy of which A. Haseloff (*Die Bauten der Hohenstaufen in Unteritalien*, Leipzig 1920) especially emphasized.

58 They stem from the superintendent in Bari, the architect Riccardo Mola, who so kindly made the results of his observations on the occasion of restoration work available to me.

59 Reproduced in Stierlin 1979, p. 162. Further examples are: (a) the sanctuary of the building complex in the center of Cairo erected by Sultan Qalāwūn between 1283 and 1294, (b) the Köşk madrasa, built in 1339, and (c) the Şaihzade mosque in Istanbul, built between 1544 and 1548 by Sinān.
Antonio Filarete, who worked from 1451 to 1465 for Francesco Sforza and was fond of geometry, took the same sketch as the basis of his design for a chapel for the Ospedale Maggiore in Milan (see the *Lexikon der Weltarchitektur*, edited by Nikolaus Pevsner with Hugh Honour and John Fleming, Munich 1987[2], p. 296).

60 *Repertorio*, p. 32ff.; Hotz 1981, p. 313ff., Figs. Z 184–186, plate 200.

61 Willemsen 1968, p. 25ff.

62 See Guido di Stefano, *Monumenti della Sicilia Normanna*, edited by Wolfgang Krönig, Palermo 1979[2], Figs. 321–325, plates CXCV–CXCVII. The question of where this form of truncated pyramid base comes from remains open: from Arab fortifications with sloping loam walling, or from the earth fills of European 'mottes'? Cf. H. Hinz, *Motte und Donjon*, Cologne/Bonn 1981, pp. IIff., 50, 76.

63 We have already considered the plan of the lower church of Santa Maria di Siponto (see p. 59, Fig. 75).

64 See Riccardo Mola, 1983, p. 331ff., Fig. 48 and plate VI.

65 The sketchbooks of Leonardo da Vinci also contain grid arrangements — for instance, in connection with design sketches for centrally planned buildings in the ratio of 3:3, in rare cases 5:5. Cf. Carlo Pedretti, *Leonardo da Vinci, Architekt*, Stuttgart/Zurich 1981, Fig. 31 (Codex Atlanticus, fol. 362 V-b). Grids as the basis for design sketches appear to have been widely used during the Renaissance. Whether they have anything to do with the Pythagorean triples described here seems questionable. Even the Egyptians used square grids for designs and structural drawings. Cf. Helmuth Gericke, *Mathematik in Antike und Orient*, Berlin/Heidelberg 1984, p. 56.

66 Cf. Werner Körte, 'Das Kastell Kaiser Friedrichs II. in Lucera', in: *25 Jahre Kaiser-Wilhelm-Gesellschaft zur Förderung der Wissenschaften*, vol. III: Die Geisteswissenschaften, Berlin 1937, pp. 51–53.

67 Huillard-Bréholles 1857, vol. V, part II, p. 912: decree of April 22, 1240 (Foggia).

68 *Repertorio*, p. 79ff.; Michele Cordaro: 'La porta di Capua', in: *Annuario dell'Istituto di storia dell'arte dell'Università di Roma*, Rome 1974–76; Hotz 1981, pp. 326ff. and 334, Fig. Z 190, plates 206, 207. Willemsen 1953. Further: Ferdinando Bologna, 'Cesaris Imperio

Regni Custodia Fio', La Porta di Capua e la "Interpretatio Imperialis" del Classicismo', in: *Nel Segno di Federico II*, Naples 1989, pp. 159–189, Figs. 1–23.

69 Muratori 1938, p. 188. — Willemsen 1953, p. 7f. and notes 5–7, p. 77.
70 Willemsen 1953, p. 5.
71 Claussen 1990, pp. 19–39.
72 Ibid., p. 298, Figs. 4 and 14.
73 Deichmann 1983, p. 73f., justifiably questions the interpretations, as does Willemsen 1953, p. 49f.
74 Willemsen 1953, p. 37.
75 In *Repertorio*, p. 71, erroneously reproduced as regular octagons.
76 Hanno Hahn and Albert Renger-Patzsch, *Hohenstaufenburgen in Süditalien*, Ingelheim am Rhein 1961, p. 34f., text reproduction 24.
77 To name only *one* example of which we happen to be aware: Theodorus of Antioch ('Master Theodorus'), a Jacobite Christian, had studied under the eminent Arab encyclopedist Kamāl ad-Dīn ibn Yūnus before he arrived at the court of Frederick II, where he met, among others, Leonardo of Pisa. Cf. Heinrich Suter, 'Beiträge zu den Beziehungen Kaiser Friedrichs II. zu den zeitgenössischen Gelehrten des Ostens und Westens, insbesondere zu dem arabischen Enzyklopädisten Kamāl ad-Dīn ibn Yūnus', in: *Abhandlungen zur Geschichte der Naturwissenschaften und der Medizin*, no. IV, Erlangen 1922, p. 8. Cf. also F. Gabrieli, 'Federico II e il mondo dell'Islam', in: *Nel Segno di Federico II*, Naples 1989, pp. 143–158, especially p. 135.
78 *Repertorio*, p. 63ff.
79 Creswell Shearer, *The Renaissance of Architecture in Southern Italy. A Study of Frederic II of Hohenstaufen and the Capua Triumphator Archway and Towers*, Cambridge 1935. — M. d'Onofrio, 'La Torre Cilindrica de Caserta Vecchia', in: *Napoli Nobilissima*, vol. VII, no. 1, 1969, p. 33.
80 *Repertorio*, p. 66, Fig. 9.
81 *Repertorio*, p. 190, and *Il Castello dell'Imperatore a Prato*, edited by Francesco Gurrieri, Prato 1975; G. Agnello, 'Il castello svevo di Prato', in: *Rinasa*, III, 1954, pp. 147–227; Hotz 1981, pp. 290 and 334, Fig. Z 168, plates 183/84.
82 Willemsen 1968, p. 154, Fig. 11 (plan after Agnello 1954).
83 *Repertorio*, p. 23ff.; Hotz 1981, p. 314f., Fig. Z 183, plate 198.
84 Willemsen 1968, p. 27, Fig. 12, and Maria Giuffrè, *Castelli e luoghi forti di Sicilia (XII–XVII secolo)*, Palermo 1980, p. 14f., Fig. 9 (ground plan).
85 Otto Lehmann-Brockhaus, *Abruzzen und Molise*, Munich 1983, p. 331, Figs. 59, 209.
86 Wollin 1935, p. 238, Fig. 113.

87 Saro Franco and Massimo Cultraro, *Guida attraverso i saloni del Castello Normanno di Adranò*, Adranò, no date.
88 *Repertorio*, p. 162ff.; Giuffrè 1980, p. 24, Figs. 23–25.
89 Krönig 1948, p. 948. It seems to me that it has not yet been demonstrated in which direction an influence may be assumed.
90 Ernst Adam, *Die Baukunst der Stauferzeit in Baden-Württemberg und im Elsaß*, Stuttgart (Aalen 1977, p. 213ff., Fig. 106. — Cf. Hotz 1981, p. 183, Fig. Z 90, plate 101.
91 Cf. also the keep of the ruins of Staufeneck near Göppingen, which is round outside and octagonal (!) inside: Adam 1977, p. 212ff., Figs. 104 and 105.
92 Christian Wilsdorf, 'Le Burgstall de Guebwiller, Les châteaux octogonaux d'Alsace et les constructions de l'empereur Frédéric II', in: *Annuaire de la Société d'Histoire des régions de Thann-Guebwiller*, 1970–72, p. 14.
93 Cf. ibid., pp. 9–16. On p. 13 is a list of octagonal fortresses in the area of the Alsace and Württemberg, all of which are destroyed, however, and date back to the thirteenth and fourteenth centuries, according to the author!
94 Regarding Margrave Dietrich von Meißen, see *Lexikon des Mittelalters*, vol. III, 1986, columns 1223/1224.
95 Regarding Heinrich von Morungen, see *Deutsche Literatur des Mittelalters — Verfasser Lexikon*, vol. 3, 1981, columns 804–815.
96 K. List, 'Die Wasserburg Lahr', in: *Burgen und Schlösser. Zeitschrift für Burgenkunde*, 1970; Adam 1977, p. 125f., Fig. 57; Hotz 1981, p. 184ff., Figs. Z 19 and Z 92, plate 102.
97 Krönig 1948, p. 949; Hotz 1981, p. 167f., Fig. Z 81, plate 83.
98 Nikolaus Flüeler, *Das große Burgenbuch der Schweiz*, Munich 1977, figures pp. 139, 180; cf. also the fortress Marschlins in Graubünden. Hotz 1981, p. 276f., Fig. Z 158, plate 174.
99 Bodiam Castle in Sussex, cf. Patrick Cormack, *Castles of Britain*, New York 1982, pp. 92 and 93; cf. also Willemsen 1968, p. 47, Fig. 28.
100 Reinhart Wolf, *Castillos, Burgen in Spanien*, Munich 1983, Fig. 17, p. 107.
101 The fortress of Krefeld-Linn must also be mentioned, where a donjon-like building was later surrounded with a rectangular wall with round corner towers. Cf. Hinz 1981, p. 50, Fig. 27. Finally, the Rocca of Imola (ca. 1502) should be mentioned here; cf. Pedretti 1981, p. 166f., Figs. 236–238 with ground-plan sketches by Leonardo da Vinci (Windsor no. 12686).
102 Domenico da Cortona and others, begun 1519. Klaus Merten, in: *Propyläen-Kunstgeschichte*, vol. 8, Berlin 1970, p. 370, Figs. 32, 373.

103 See among others Ernst Kühnel, 'Islamische Einflüsse in der romanischen Kunst', in: *Sitzungs-Berichte der Kunstgeschichtlichen Gesellschaft Berlin*, Nov. 11, 1927.
104 See E. Merra, *Castel del Monte*, Molfetta 1964³, p. 32, and E. Swinburne, *Viaggio nelle due Sicilie negli anni 1777 e 1778*.
105 I have not taken into account the assumptions of Aldo Tavolaro ('Il Sole Architetto a Castel del Monte', Appendix to the book *Castel del Monte* by C. A. Willemsen, Bari 1984), as they are difficult to prove, because neither the height of the wall crest of Castel del Monte nor the level of the inner court are known exactly. Both dimensions are nevertheless carelessly used for a theory of the angle of incidence of the sun on a certain day of the year. Riccardo Mola, *Restauri in Puglia (1971–1983)* II, Fasano (Brindisi) 1983, p. 20ff.
106 See Chierici 1934, plates I, II, and VI–VII. According to Chierici's sectional drawings (vols. VI–VII), the outer walls are recessed less than the walls toward the inner court.
107 Cf. Merra 1964, p. 31.
108 On this point see note 104.
109 Cf. in this regard the intarsia patterns in Krönig 1978, plate CLIX (cathedrals of Teano and Capua).
110 Etienne Gruyon and H. Eugene Stanley, 1991.
111 Émile Bertaux, 'Castel del Monte et les architects français de l'empereur Frédéric II', in: *Comptes-rendus des séances de L'Académie des Inscriptions et Belles Lettres*, Paris, August 1897, p. 432ff.
112 Ibid., p. 434. — Bertaux' argumentation on the basis of pure comparison of types and the assumption of a direct and lively connection of the Cistercian lodges with the focal points of Gothic in central Europe put him in the difficult situation of having to state that, stylistically, the southern Italian forms of capitals lagged about half a century behind the 'classical' development of Gothic. This is not surprising, however, if one acknowledges the fact that the lodges worked with pattern books from their French homeland and thus did not take into consideration the further developments in the artistic centers of Gothic — particularly as they had at their new workplaces new and other problems to master, with which Frederick II and his architects confronted them. Apt here is also the observation (C. A. Willemsen, *Castel del Monte*, Insel-Bücherei no. 619, Frankfurt a.M. 1982²) that the tunnel vaulting of Castel del Monte had long been replaced by ribbed vaulting in French buildings of the same period.
113 Bertaux himself has to concede that the roof design of Castel del Monte does not conform with 'French' tradition: cf. Émile Bertaux, *L'Art dans l'Italie Méridionale de la fin de l'Empire romain à conquête de Charles d'Anjou*, vol. 2, Paris 1904, p. 743f.
114 Cf. Otto von Simson, *Die gotische Kathedrale*, Darmstadt 1982⁴.
115 Giosuè Musca, *Castel del Monte, Il Reale e l'Immaginario*, Bari 1981, p. 51, Fig. 24; in addition Carl Arnold Willemsen, *Apulien, Kathedralen und Kastelle. Ein Kunstführer durch das normannisch-staufische Apulien*, Cologne 1973², Fig. 63.
116 Bernhard Schweitzer, *J. G. Herders 'Plastik' und die Entstehung der neueren Kunstwissenschaft*, Leipzig 1948.
117 Hermann Weyl, 'Symmetrie', in: *Wissenschaft und Kultur*, vol. II, Boston/Basel/Stuttgart 1981², p. 122.
118 Cf. Shearer 1935. In addition: M. L. Testi Cristiani, *Nicola Pisano architetto e scultore*, Pisa 1987; see also Bologna 1989, pp. 159–189, Figs. 1–23.
119 With regard to this and what follows, cf. von Simson 1982, especially p. 53ff.
120 Moritz Cantor, *Vorlesungen über die Geschichte der Mathematik*, vol. 1, Leipzig 1907³, p. 102ff.
121 Andreas Speiser, *Die mathematische Denkweise* (Wissenschaft und Kultur, 1), Basel 1945², p. 17f. Concerning the architects of the Hagia Sophia, Anthemios and Isidorus, cf. G. Downey, 'Byzantine Architects, Their Training and Methods', in: *Byzantion, Revue Internationale des Études Byzantines*, vol. 18, 1948, pp. 99–118, especially p. 112ff.
122 H. R. Hahnloser, *Villard d'Honnecourt: Kritische Gesamtausgabe des Bauhüttenbuches*, Ms. fr. 19093 of the Bibliothèque Nationale de Paris, published by Anton Schroll & Co. in Vienna 1935. Cf. Cord Meckseper, 'Über die Fünfeckkonstruktion bei Villard de Honnecourt und im späten Mittelalter', in: *Architectura. Zeitschrift für Geschichte der Baukunst*, vol. 13, 1, Munich 1983, p. 31ff.
123 Lon R. Shelby, *Gothic Design Techniques. The Fifteenth Century Design Booklets of Mathes Roriczer and Hanns Schmuttermayer*, London/Amsterdam 1977.
124 The term *kanon* originated in Greek architecture, where it signified a measuring pole made of wood; cf. Argyres Petronotis, *Bauritzlinien und andere Aufschnürungen am Unterbau griechischer Bauwerke in der Archaik und Klassik. Eine Studie zur Baukunst und -technik der Hellenen* (published by the author), Munich 1968, p. 22f.
125 Cf. for instance the examples in the stimulating contribution by Meckseper 1970, p. 213. The examples above the illustration of Castel del Monte do *not* represent preliminary stages of this building.
126 Concerning the number 8: F. C. Endres and A. Schimmel: *Das Mysterium der Zahl*. Diederichs Gelbe Reihe, 2nd edn. 1985.

127 Cf. Pedretti 1981, p. 160, Fig. 230 (drawing after Vitruvius, ca. 1490, Venice, Galleria dell'Accademia).

128 Filippo Franciosi, 'L'Irrazionalità nella Matematica Greca Arcaica', in: *Bollettino dell'Istituto di Filologia Greca dell'Università di Padova*, Suppl. 2, Rome 1977. Cf. also: Seyyed Hossein Nasr, *An Introduction to Islamic Cosmological Doctrines*, Cambridge, Mass. 1964, p. 47ff.

129 The Carolingian Beda manuscript at the Benedictine abbey of St. Maximin in Trier shows how vivid this conception of the meaning of the square was as an expression not only of the four compass directions but at the same time of the order of the world in the early Middle Ages. This manuscript testifies to the rise of Western culture on Carolingian ground as a conglomeration of classical heritage and Christian thought. On one parchment of this manuscript from the eleventh century, under f99, there is a compass rose with a diagram of the earth and the notation: "The world has four corners and is divided into four parts" (see cat. of the special exhibition *Karolingische Handschriften, Beda-Handschriften und Frühdrucke* in the Stadtbibliothek Trier from May 12 to November 3, 1990, published by the Cultural Foundation of the Federal States and the Stadtbibliothek Trier, Trier 1990, p. 64, no. 33, Fig. 99 on p. 63).

130 Derek J. de Solla Price, 'Geometrical and Scientific Talismans and Symbolisms', in: *Science since Babylon*, New Haven/London 1976, p. 71 f.

131 John V. Nobel and Derek J. de Solla Price, 'The Water Clock in the Tower of the Winds', in: *American Journal of Archaeology*, 1968, pp. 345–355. Rolf C. A. Rottländer/Werner Heinz/Wilfried Neumaier, 'Untersuchungen am Turm der Winde in Athen', in: *Jahreshefte des Österreichischen Archäologischen Institutes*, vol. LIX, Vienna 1989, pp. 55–92. It may be of interest in this connection that Frederick II received as a gift from Sulṭān al-Aṣraf of Damascus a tent in which pictures of the sun and the moon followed their course and told the hours of day and night exactly. These were systems described repeatedly in Arab sources. Cf. E. Wiedemann, 'Miszellen. Fragen aus dem Gebiet der Naturwissenschaften, gestellt von Friedrich II., dem Hohenstaufer', in: *Archiv für Kulturgeschichte*, vol XI, no. 4, Leipzig/Berlin 1914, p. 485.

132 Richard Krautheimer, 'Introduction to an Iconography of Medieval Architecture', in: *Journal of the Warburg and Courtauld Institutes*, vol. V, 1942, p. 1ff., especially p. 23; Gioacchino de Angelis d'Ossat, 'Sugli edifici ottogonali a cupola nell'Antichità e nel Medioevo', in: *Atti del 10 Congresso Nazionale di Storia dell'Architettura*, Florence 1938, p. 13ff.

133 Franz Josef Dölger, 'Zur Symbolik des altchristlichen Taufhauses. Das Oktogon und die Symbolik der Achtzahl', in: *Antike und Christentum*, Münster/Westf. 1933, p. 153ff. — Cf. also Paul A. Underwood, 'The Fountain of Life in Manuscripts of the Gospels', in: *Dumbarton Oaks Papers*, no. 5, Cambridge, Mass. 1950, p. 41ff., especially Fig. 23 (baptistry of the Lateran basilica) and Fig. 26 (8 columns!). See in addition Appendix B: 'The Six and the Eight in Baptisteries and Fonts: Archaeological Evidence', p. 131ff.

134 Alexandre Papadopoulo, *Islam and Muslim Art*, London 1980, p. 247.

135 For example, in the mausoleum of the Samanids in Buhārā (here p. 122f. and note 147) or the mausoleum of Ġiyāṯ ad-Dīn Tuġluq (1321–1325), Fort Tuġluqabād Delhi: A square ground plan is changed via the octagon and hexagon into the circular base of the cupola.

136 Codex Medicus Dioscurides 'De Materia Medica', Vienna, Austrian National Library, Cod. Med. Gr. 1, fol. 6V; Kurt Weitzmann, *Spätantike und frühchristliche Buchmalerei*, Munich 1977, p. 60f., Fig. 15 on p. 60.

137 The question of whether this symbology arose in East Asia independent of the West or the Near East cannot be dealt with here. The close ties that existed between the Islamic and the Chinese world during the Tang Dynasty and continued up to the Mongol period would appear to indicate a connection. The Sassanids sent legations to Chang'an in the seventh century. The influence of Iran on art and handcraft in China during the seventh and eighth centuries is noticeable; cf. Jacques Gernet, *Le Monde Chinois*, Paris 1972 (German edition *Die chinesische Welt*, Frankfurt a. M. 1979, p. 238ff.). The symbolism of the octagon seems to be deeply rooted in human imagination, however; its foundations are no doubt to be found in the world's four directions of the compass or of movement. Cf. the numinous nature of the octagon or the eight-pointed star also in Central American – pre-Columbian art.

138 This is a coffer of 5.94 m², recessed approximately 1.80 m. Above the basic square frame, which is about 50 cm deep, lies the eight-pointed star formed from two crossed squares with a depth of another 57 cm. Within the central octagon vaults a circular dome with a depth of 72.2 cm and a diameter of 3.2 m. The individual stages follow the familiar sequence: square (earth) — octagon (two crossed squares) — circle (heaven). A more significant location for the symbolism of the transcendence from the earth to heaven, or of the emperor's heavenly power is hardly conceivable. Here on p. 119, Fig. 166e, from cat. *Collection of Fine Arts in China*, vol. 1 'Palace Architecture' (Chinese), Beijing 1987, plate 23.

139 *Palaces of the Forbidden City*, compiled by Yu Zhuoyun, Hong Kong 1984, p. 254, Fig. 357.

140 Cat. *Collection of Fine Arts* 1987, p. 114, Fig. 111.
141 Zhuoyun 1984, p. 256, Fig. 357.
142 *Historic Chinese Architecture*, compiled by the Department of Architecture, Qinghua University, Beijing 1985, p. 71.
143 Cf. Dietrich Seckel, 'Taigenkyû, das Heiligtum des Yuiitsu-Shintô. Eine Studie zur Symbolik und Geschichte der japanischen Architektur', in: *Monumenta Nipponica*, 6, Tokyo 1943, pp. 52–85.
144 Burchard Brentjes, *Kunst des Islam. Mittelasien*, Leipzig 1979, 1982², Fig. p. 24.
145 Alistair Northedge and Robin Falkner, 'Survey Season at Sāmarrā 1986', in: *Iraq*, 49/1987, pp. 143–173.
146 Keppel Archibald Cameron Creswell, *A Short Account of Early Muslim Architecture*, Harmondsworth 1958, p. 286ff., Fig. 59. — Cf. *Propyläen-Kunstgeschichte*, vol. 4, Berlin 1973, p. 218, Fig. 35.
147 Cf. Papadopoulo 1980, p. 527 and Fig. 868 on p. 528. Further: *Propyläen-Kunstgeschichte*, vol. 4, Fig. 142, plate XXVI, and p. 234, Fig. 40; M. S. Bulatov, *Geometrische Harmonisierung in der Architektur Zentralasiens im 9.–15. Jahrhundert* (in Russian), Moscow 1988², pp. 97–104, Figs. 40–47. Bulatov discusses the proportions of central Asian Islamic buildings on the basis of an anonymous Persian translation of the geometric treatise after Abū l-Wafā' al-Būzağānī (940–998): 'On That Which Artisans Must Know About Geometric Constructions' (The manuscript of the treatise is in the Bibliothèque Nationale, Paris/Persian manuscripts no. 169).
148 M. S. Bulatov devoted a monograph of his own to the Samanid mausoleum: *Das Mausoleum der Sāmāniden, Perle der Architektur Mittelasiens* (in Russian), Tashkent 1976. — Cf. in addition *Architecture of the Islamic World*, edited by George Michell, London 1978, Fig. p. 250 and plate 17. Finally: Yvonne Dold-Samplonius, 'The Volume of Domes in Arabic Mathematics', in: *Vestigia Mathematica. Studies in medieval and early modern mathematics in honour of H. L. L. Busard*, Amsterdam 1993, p. 93ff.
149 Bulatov 1988², p. 121ff., Figs. 64–68. Further mausoleums with square plans, octagonal upper stories, and cupolas with an eight-part arrangement or a hexagonal pyramid on top dating from the tenth to the fifteenth century are listed by Bulatov in the fifth chapter of his work (p. 96ff.):
  a) Mausoleum of 'Arab-Ātā in Tīm (977–978): p. 105ff., Figs. 48, 49.
    Square plan; upper story with eight-sided cupola.
  b) Mausoleum of Čaugāndar-Bābā in Turkmenistan (11th–12th cent.): p. 116ff., Fig. 57.
    Lower story square, upper story octagonal with eight-sided pyramid.
  c) Mausoleum of Yusuf ibn Kutaiyir in Nahǧawăn, Aḏerbaiǧān (1162): Burchard Brentjes/L. Bretanizki/B. Weimarn, *Die Kunst Aserbaidschans vom 4. bis zum 8. Jahrhundert*, Leipzig/Weinheim 1988, p. 61f. with fig., plates 23, 24.
    Octagonal prism with eight-sided pyramid. The architect was 'Aǧamī ibn Abū Bakr an-Nahǧawānī.
  d) Mausoleum of Abū l-Muẓaffar Takaš in Kunya Urgenç (12th–13th cent.): p. 128ff., Figs. 71–73.
    Square lower story, octagonal mezzanine. Transition to a decahexagon and a round cupola.
    Following the Mongolian invasion:
  e) Mausoleum Mazlūm-Ḥan-Sulūw on the grounds of the necropolis of Mazdaqān (end of 13th cent.): p. 132ff., Figs. 74, 75.
    Complex with a square main room, which develops into an octagonal upper story and a cupola resting on it, also octagonal. A smaller side room shows the same geometric arrangement, including the cupola. A square connecting room of the staircase has a similar structure.
  f) Mausoleum of Quṭam ibn 'Abbās in the building complex Šāh i Zinde (1334–1335): p. 145ff., Fig. 82. G. A. Pugačenkova, *Chefs-d'œuvre d'architecture de l'Asie centrale XIV–XV siècle*, Paris 1981, p. 82.
    Main building with a square plan and an octagonal upper story. It is topped by an octagonal cupola. The annex displays the same structure.
  g) Mausoleum of Boyān Qūlī-Ḥān in Buḫārā (1358): p. 147ff., Figs. 83, 84.
    Square lower story and octagonal upper story with round cupola.
  h) Nameless mausoleum in the Šāh-i Zinde building complex in Samarqand (1360–1361): p. 149ff., Figs. 85, 86. Pugačenkova 1981, p. 52, Fig. 40d.
    Square main story, octagonal mezzanine and cupola.
  i) Mausoleum of Šādī Mulk Āqā in the Šāh-i Zinde building complex in Samarqand (1371–1372): p. 157ff., Figs. 90–92, *Propyläen-Kunstgeschichte*, vol. 4, Fig. 314, Pugačenkova 1981, p. 49, Fig. 37a, p. 68, Fig. 3, p. 82ff., especially p. 84, Figs. 4, 5.
    Square lower story with octagonal upper story and decahexagon lying above it, in which the circle of the cupola is inscribed. The cupola is decorated with an eight-pointed star. Cf. here p. 128, Fig. 185.
  j) Mausoleum of Bībī-Ḥanum in Samarqand (beg. 15th cent.): pp. 181, 184, Fig. 114; Pugačenkova 1981, p. 52, Fig. 41a, p. 110f.
    Main building shaped as octagonal prism with square interior. Over it a high cylindrical tambour with a spherical-conical cupola.
  k) Mausoelum of Šaiḫ Dursun in Aḏerbaiǧān (1403–1404): p. 184f., Fig. 115.

The mausoleum of Šaiḫ Dursun near Aqsu (Aḏerbaiġān) belongs to the same period as the mausoleums of Tamerlane, Gūr-e Mīr, his wife Bībī-Ḥanum, and Tumān Āqā in Samarqand. The monument has the shape of an octagonal prism with relatively thin walls and an octagonal pyramid-shaped tower. The building's proportions are shown in the plan (here p. 155, Fig. 224).

l) Mausoleum of Gūr-e Mīr in Samarqand (1404): p. 176ff., Figs. 107–111. Pugačenkova 1981, p. 52, Fig. 40a, p. 112f.
Octagonal prism. Over it a cylindrical tambour and cupola.

m) Ǧauhar Šād in Kušān (Afġānistān, 1440): Pugačenkova 1981, p. 52, Fig. 41b, pp. 144–147.
Octagonal foundation with cylindrical tambour and cupola.

n) Dīwān-Ḫāne in Bākū (15th cent.): pp. 200, 203, Fig. 35.
Octagonal foundation with gallery on pillars.

o) Octagonal mausoleum in Šāh-i Zinde in Samarqand (15th cent.): Pugačenkova 1981, p. 52, Fig. 40f., pp. 82 and 92, Figs. 21 and 22.
Octagonal prism with flat cupola.

p) Rabīʿa Sulṭān Begum in Turkestan (15th cent.): Pugačenkova 1981, p. 52, Fig. 40c.

150 Michell 1978, p. 268.
151 Ibid., p. 239.
152 Ibid., p. 272.
153 Stierlin 1979, p. 247; John D. Hoag, *Islam*, Stuttgart 1986, p. 163, Fig. 280.
154 Alund Nabi Khan, *Multan — History and Architecture*, Institute of Islamic History, Culture and Civilisation, Islamic University, Islamabad 1983.
155 The mausoleum of Ismāʿīl the Samanid (see here p. 123, Fig. 172) also displays round corner towers, albeit on a square building.
156 Khan 1983, p. 217, plate I; Tonny Rosiny, *Pakistan*, Cologne 1983, p. 279, Fig. 41.
157 Khan 1983, p. 223, Fig. 34.
158 Ibid., p. 236ff., Fig. 49 and plate II.
159 Klaus Fischer, Michael Jansen, and Jan Pieper, *Architektur des Indischen Subkontinents*, Darmstadt 1987, p. 197, Fig. 209.
160 The architect of the empress Theodora, wife of Justinian, who had also built the Church of the Apostles at Constantinople, was praised for being able to draw (scratch) the linear projection of the stereometric body of this church on level ground. Stereometric bodies can mean only the spatial elements — not a model — which had to be scarified into the ground of the building site two-dimensionally. The vaulted ribs of the two stereometric cupola rooms of Córdoba, for example, were first sketched linearly as two squares turned toward one another by 45°, or as a regular eight-pointed star, and projected from there in the course of building into the stereometric space of the cupola. Cf. Petronotis 1968, p. 33. Such techniques were most certainly maintained over many centuries. Cf. on this point in general also Petronotis, p. 7f.
161 Cf. the same view on this held by Stierlin 1979.
162 Hoag 1986, p. 46, Fig. 74.
163 Hans Erich Kubach, 'Romanik', in: *Weltgeschichte der Architektur*, Stuttgart 1986, p. 152, Fig. 243.
164 Galine Anatol'evna Pugačenkova, *Puti razvitija architektury Južnogo Turkmenistana pory rabovladenija i feodalizma*, Moscow 1958, p. 363. The ceiling decoration in the cupola of the mausoleum of Šādī Mulk Āqā in Samarqand (1371) is related. *Propyläen-Kunstgeschichte*, vol. 4, Fig. 314 (here p. 128, Fig. 185).
165 Bulatov 1988[2], p. 177ff., Fig. 91 (see here note 149 under i).
166 B. Lewis, *The World of Islam*, London 1976, p. 244, Fig. 17; Burchard Brentjes, 'Das 'Ur-Mandala' (?) von Daschly-3', in: *Iranica Antiqua*, vol. 18, 1983, p. 39, plate Ib; Titus Burckhardt, *Die maurische Kultur in Spanien*, Munich 1980, p. 181, Fig. 62 (in color).
167 *Propyläen-Kunstgeschichte*, vol. 6, p. 165f. Fig. 140.
168 Cf. Papadopoulo 1980, p. 288; G.D. Lowry, 'Humayun's Tomb: Form, Function and Meaning in Early Mughal Architecture', in: *Muqarnas. An Annual on Islamic Art and Architecture*, Leiden 1987, vol. 4, pp. 133–148. Lowry regards the six-pointed star decoration, visible in important places, as characteristic for the symbology of the Mogul dynasty. Thus, it is all the more striking that the octagonal (!) central room beneath the cupola, in the middle of which Humāyūn's sarcophagus stands, is dominated by the eight-pointed star — just as in the Tāǧ Maḥall (p. 131, Fig. 189)!
169 Issem El-Said and Ayşe Parman, *Geometric Concepts in Islamic Art*, London 1976, pp. 7–49, especially Fig. 28, p. 37. Cf. also A.J. Lee, 'Islamic Star Patterns', in: *Muqarnas. An Annual on Islamic Art and Architecture*, Leiden 1987, vol. 4, pp. 182–197. Corresponding examples are found in the standard work on surface ornamentation by Owen Jones, *The Grammar of Ornament*, folio edition, London 1856. In spite of its opulence it was not able to cover the ebullient geometric fantasy of Islamic ornamental art entirely. A treasure of this ornamental art is concealed in the work by Gerd Schneider, *Geometrische Bauornamente der Seldschuken in Kleinasien*, Wiesbaden 1980. Cf. also Kurt and Hanns Erdmann, *Das anatolische Karavansaray des 13. Jahrhunderts*, part III: 'Die Ornamente', Berlin 1976.
170 André Paccard, *Le Maroc et l'artisanat traditionnel islamique dans l'architecture*, Édition Atelier 74,

2 vols., Annecy 1983⁴, vol. 1, p. 210. Cf. Ernst H. Gombrich, *Il senso dell'ordine*, Turin 1984, p. 146, Fig. 97. Further: Edith Müller, *Gruppentheoretische und strukturanalytische Untersuchungen der maurischen Ornamente aus der Alhambra in Granada* (Dissertation), Zurich 1944, plate 8, Fig. 5.
171 Peter Murray, *Die Architektur der Renaissance in Italien*, Stuttgart 1980, p. 84ff., Fig. 66 on p. 87 (manuscript M S.B., ca. 1498). — Cf. also Pedretti 1980, p. 246, Fig. 360 (Codex Atlanticus, folio 271 V–d). There, next to a 'false' eight-pointed star over a square, stands the 'right' one with especially accentuated points. The construction common to the Islamic area does not appear to be known.
172 Cf. Gerda Soergel, *Untersuchungen über den theoretischen Architekturentwurf von 1450–1550 in Italien*, Munich 1958, pp. 59ff. and 64.
173 Ibid., p. 95f. and Fig. 23 on p. 127.
174 And furthermore also in Santa Sindone; cf. Guarino Guarini, *Architettura civile del padre di Guarino Guarini... opera postuma...*, Turin 1737, plates 4–6. — Mario Passanti, *Nel Mondo Magico di Guarino Guarini*, Turin 1963.
175 Norbert Huse, in: *Propyläen-Kunstgeschichte*, vol. 9, Berlin 1970, p. 219f., Figs. 4 and 229b; Frederico A. Arborio Mella, *Gli Arabi e l'Islam, Storia, Civiltà, Cultura*, Milan 1981, p. 280 with figure.
176 Cf. George Kubler and Martin Sebastian Soria, *Art and Architecture in Spain and Portugal and their American Dominions*, Harmondsworth 1959, p. 29.
177 The book by Mathias Untermann, *Der Zentralbau im Mittelalter*, Darmstadt 1989, contains a valuable collection of centrally planned religious buildings from the period of the Latin church from the seventh to the fifteenth century. A related discussion of the overall history of centrally planned buildings is still not available, however.
178 Cf. Creswell 1958, pp. 17–42. In addition Papadopoulo 1980, pp. 239–243, and Doron Chen, 'The Design of the Dome of the Rock in Jerusalem', in: *Palestine Exploration Quarterly*, 112, 1980, pp. 41–50.
179 Klaus Brisch, in: *Propyläen-Kunstgeschichte*, vol. 4, p. 143, and Robert Hillenbrand, ibid., p. 154.
180 Ernst Kantorowicz, *Kaiser Friedrich der Zweite*, Berlin 1927, vol. 1, p. 177, supplementary vol., p. 68. Report by Makrizi in M. Amari, *Biblioteca Arabo-Sicula*, two vols., Turin/Rome 1880f., p. 521 (Bibliography) and vol. II, p. 265. — The Siedler-Verlag published a German translation of the work *Frederick II, A Medieval Emperor*, London 1988, by David Abulafia in 1991. It is a subjective representation of Frederick II of Hohenstaufen, directed against both his importance and his historiographer Ernst Kantorowicz. The architectural history is hardly considered in the book. It contains nothing noteworthy about Castel del Monte except for a disparaging characterization: "Castel del Monte is best described as a hunting-box. The interior of the building is remarkable for its ribbed vaults, but it is a singularly unexciting place inside; once you have seen one room you have seen them all" (ibid., p. 288).
181 *Revue Archéologique*, third series, vol. XII, 1888, pp. 14–23.
182 Papadopoulo 1980, p. 239f.
183 Ecochard describes a large number of other churches and cathedrals — primarily in the Byzantine region — the dimensions of which are identical to those of the buildings described here or equal to half of them. It may thus be concluded that these dimensions were 'canonized' within the framework of the eight-pointed star as leitmotif.
184 Cf. in this regard: Franz Carl Endres and Annemarie Schimmel, *Das Mysterium der Zahl*, Cologne 1984³, pp. 172–179.
185 See von Simson 1982, p. 33ff. — P. Frankl, 'The Secret of the Medieval Masons', in: *The Art Bulletin*, vol. XXVII, no. 1, March 1945, pp. 46–64. Further: J. S. Ackermann, "Ars sine scientia nihil est.' Gothic Theory of Architecture at the Cathedral of Milan', in: *The Art Bulletin*, vol. XXXI, no. 1, March 1949, pp. 84–111. Cf. also E. Violett-le-Duc, *Entretiens sur l'Architecture*, I, Paris 1863, pp. 284–287. 'Ars' refers to practice, 'scientia' to scientific geometry.
186 Cf. *Propyläen-Kunstgeschichte*, vol. 6, Figs. 174 and 178.
187 Shelby 1977, p. 4.
188 Hahnloser 1935.
189 Ibid., plate 38.
190 Michell 1978, p. 114.
191 See Hahnloser 1935, pp. 92 and 196.
192 Ibid., plate 39.
193 See note 123. Shelby 1977, p. 85ff., Frankl 1945, p. 46ff., and K. Hecht, 'Maß und Zahl in der gotischen Baukunst', in: *Abhandlungen in der Braunschweigischen Wissenschaftlichen Gesellschaft*, Hildesheim/New York 1979, p. 171ff., Figs. 24 and 25.
194 Shelby 1977, p. 108ff., Figs. 19 and 20; cf. also Frankl 1945, p. 50ff. and Fig. 1.
195 Cf. H. Koepf, *Die gotischen Planrisse der Wiener Sammlungen*, Vienna/Cologne/Graz 1969, Figs. 341, 342, 347, 365, 464. Cf. also p. 116, this volume, Fig. 163.
196 Shelby 1977, p. 67.
197 For this reason, Lon R. Shelby formulates it as follows: "This nonmathematical technique I have labeled *constructive geometry*..." in his contribution 'The Geometrical Knowledge of Medieval Master

Masons', in: *Speculum. A Journal of Medieval Studies*, vol. XLVII, Cambridge, Mass. 1972, p. 420. In the same work he also writes (p. 407): "Since neither Villard nor Magister 2 gives evidence of using either the astrolabe or the surveyor's quadrant, one suspects that they were not abreast of the advances in surveying which had come to Europe *with the introduction of Arabic learning and mathematical instruments*" (my italics).

198 Von Simson 1982, p. 293.

199 Frankl (1945, p. 46ff.) covered this in great detail; cf. also Ackermann 1949, pp. 84–111.

200 Frankl 1945, pp. 46–51 ("AD QUADRATUM ET ED TRIANGULUM"); further: Ackermann 1949, pp. 91, under [3], and 92.

201 Frankl 1945, p. 47ff., and Shelby 1977, pp. 46–61.

202 Hecht 1979. In addition: Albrecht Kottmann, *Fünftausend Jahre Messen und Bauen*, Stuttgart 1981, p. 94, and by the same author, 'Zusammenhänge zwischen antiken Längenmaßen', in: Dieter Ahrens and Rolf C. A. Rottländer (eds), Ordo et Mensura, I. Interdisciplinary Congress of Historical Metrology, September 7–10, 1989, at the Municipal Museum Simeonstift Trier, Sankt Katherinen (unpublished MS). Cf. there the essays by Rolf C. A. Rottländer, 'Bemerkungen zur Erforschung alter Maßstäbe', p. 47ff., and 'Die mathematische Behandlung aufgemessener Längen zur Rückgewinnung der alten Maßeinheit', p. 52ff.

203 Hecht 1979, p. 285: "So verstanden war es für den gotischen Architekten eine Auszeichnung, als 'Meister der Geometrie' zu gelten — der *Geometrie*, nicht der *Proportionsgeometrie*." Hecht equates here the term 'proportional geometry' with the theories of proportion that he opposes. This does not change the fact, however, that every type of geometry is concerned with proportions, i.e., with relations between distances or angles. M. Roriczer's directions for constructing the pinnacles (*Fialen*) dealt only with proportions, no longer with dimensions, once the basic size had been fixed. See Frankl 1945, p. 49f., Fig. 1, and Shelby 1977, p. 82f.

204 Hecht 1979, p. 172.

205 Cf. here also Shelby 1977, p. 71.

206 See François Bucher, 'Design in Gothic Architecture. A Preliminary Assessment', in: *Journal of the Society of Architectural Historians*, vol. XXVII, March 1968, p. 70.

207 Ibid., pp. 49–71. The richest collection of Gothic ground plans is found in Vienna. It was published by Hans Koepf in 1969.

208 Regarding the sequence of steps he developed as well as the space of time examined, see Bucher 1968, p. 55ff.

209 Cartes et Plans, Reg. Ge B 1118.

210 David Friedman, *Florentine New Towns. Urban Design in the late Middle Ages*, Cambridge (Mass.)/London 1988, p. 138 ("The result is a visual display of proportional values that is very suggestive of the configuration adopted for the new-town plans"), see also p. 117ff., Figs. 80, 81, and 83.

211 Regarding the astrolabe, see ibid., pp. 138–140, Figs. 85 and 86, as well as David A. King, 'Astronomie im mittelalterlichen Jemen', in: *Jemen*, edited by Werner Daum, Innsbruck and Frankfurt a. M. 1987, p. 298f. Cf. also Josef Frank, 'Die Verwendung des Astrolabs nach al-Chwaritzmi', in: *Abhandlungen zur Geschichte der Naturwissenschaften und der Medizin*, no. III, Erlangen 1922.

212 Weyl 1981, p. 107.

213 Cf. the recently published contribution by Max Koecher, 'Castel del Monte und das Oktogon', in: *Miscellanea Mathematica*, Berlin/Heidelberg 1991, pp. 221–233.

214 According to the important work *La Carpinteria de lo Blanco*, Lectura dibujada del primer manuscrito de Diego López de Arenas par Enrique Nuere, Ministero de Cultura Madrid 1985, Deposito legal: M 3294. 1985, these centers are called 'sinos'. Between 1613 and 1619, Diego López de Arenas revised a manuscript concerning the architecture of Fray Andrés de San Miguel, a sixteenth-century Spanish Carmelite friar. Fray Andrés de San Miguel knew the traditions of Arabic wooden architecture that had been passed down directly and was a master of them. These principles of building with wood correspond entirely to those of ornamental design. See in particular the figure on p. 36, 'estrella de ocho puntas'.

215 In the notation by J. M. Montesinos-Amilibia, *Classical Tesselations and Three-Manifolds*, Berlin/Heidelberg 1987.

216 The height of the building — particularly following the temporary decay of Castel del Monte — can hardly be determined now. The question of whether there may have been astronomical relationships that were connected with the height of the structure must therefore remain open.

217 Shelby 1972, p. 407.

218 Cf. also the 'Gesetz ähnlicher Figuren' by August Thiersch: *Die Proportionen in der Architektur* (Handbuch der Architektur, IV/I), Leipzig 1904, and Frankl 1945, p. 55.

219 Abū l-Wafā' al-Būzağānī, born 940 in Būzağān, northeast Persia, died 998 in Bağdād. See Fuat Sezgin, *Geschichte des Arabischen Schrifttums*, vol. V, Leiden 1974, p. 321f. According to M. F. Woepcke ('Recherches sur l'histoire des sciences mathématiques chez les Orientaux d'après des traités inédits Arabes et Per-

sans', in: *Journal Asiatique*, cinquième série, vol. V, Paris 1855, pp. 218-256 and 309-359), the Persian manuscript (Bibliothèque Nationale de Paris, Persian manuscript no. 169) is a pupil's record of lessons by Abū l-Wafā' al-Būzağānī, not a work from his own hand. Parts of a manuscript of this treatise written in his own hand may be recognized in the manuscript of the Biblioteca Ambrosiana in Milan: Josef Hammer-Purgstall, *Codices Arabicos Persicos Turcicos*, Vienna 1812, p. 44, no. 68. This is only a fragment of the book, however. See Heinrich Suter, 'Beiträge zur Geschichte der Mathematik bei den Griechen und Arabern', in: *Abhandlungen zur Geschichte der Naturwissenschaften und der Medizin*, edited by Oskar Schulz, no. 4, Erlangen 1922, pp. 94-109. Still another manuscript is said to be in the Hagia Sophia in Istanbul. Concerning connections between Abū l-Wafā' al-Būzağānī and Fibonacci, see Woepcke 1855, pp. 221 and 226; concerning the life of Abū l-Wafā' al-Būzağānī, ibid., p. 243 ff.

220 Montesinos-Amilibia 1987, p. 228.
221 See Finley 1986, p. 48 f.
222 See, for instance, Hanno Hahn, *Die frühe Kirchenbaukunst der Zisterzienser*, Berlin 1957, Fig. 131 (Pontigny).
223 Bulatov 1988[2], p. 76 f.
224 Kottmann 1981, p. 96 ff.
225 Consider the discussion with Fibonacci and his invitation to revise the 'liber quadratorum'; cf. Vogel 1971.
226 Cf. Willemsen 1953.
227 *Constitutiones regum regni utriusque Siciliae...*, edited by Cajetanus Carcani, Naples 1786, p. 327 f.; Huillard-Bréholles 1852 f., vol. V, part II, p. 697; Eduard Sthamer, *Dokumente zur Geschichte der Kastellbauten Kaiser Friedrichs II....*, part II, 1926, p. 62, no. 734: "Cum pro castro quod apud Sanctam Mariam de Monte fieri volumus, per te licet de tua jurisdictione non sit, instanter fieri velimus actractum, fidelitati tue precipiendo mandamus quatenus actractum ipsum in calce, lapidibus et omnibus aliis oportunis fieri facias sine mora. Significaturus nobis frequenter quid inde duxeris faciendum..."
Dankwart Leistikow (Leistikow 1994, pp. 205-213) has made a careful study of the text of this mandate. The main subject of the discussion is the correct and adequate translation of the term 'actractus'. It is relatively clear — particularly in view of reference made by Heinz Erich Stine, a medievalist from Cologne, to English sources — that the disputed word 'actractus' should certainly be regarded as meaning 'provision of building material'. Even if its interpretation as a type of level composition 'drawing floor' is thus invalidated, this does not mean that a building with such exact basic dimensions could have been constructed without precise measurement of the building site. Methods of surveying based on the principles of Greek geometry, which were known to the Arab surveyors (see here, among others, note 197), provided the necessary techniques. With regard to dating, see also the recent publication by Gunther Wolf, 'Überlegungen zum Gründungsdatum von Castel del Monte', in: *Kaiser Friedrich II: – Manuskripte und Veröffentlichungen*, pp. 113-116. Heidelberg 1997.

228 See Bucher 1968.
229 Cantor 1907, p. 106. See also Salomon Gandz, 'Die Harpedonapten oder Seilspanner und Seilknüpfer', in: *Quellen und Studien zur Geschichte der Mathematik, Astronomie und Physik*, sect. B: Studien, vol. 1, Berlin 1931.
230 Petronotis 1968, p. 9.
231 Cantor 1907, p. 105 f.; Petronotis 1968, p. 8 f.
232 See von Simson 1982, p. 293.
233 Shelby 1977, p. 114 f.
234 Bulatov 1988[2], p. 175.
235 Cf. also Lee 1987, p. 184.
236 See the analyses by Bulatov 1988[2], p. 175.
237 Of interest to mathematicians, but perhaps also to architects, is the fact that all distances that occur in this extensive construction can be uniformly represented as a product of the power $a^k \, h^m 2^{n/2}$ with the integers k, m, and n.
238 Hahn 1957, p. 66 ff., in particular p. 69, note 212. Cf. also Fritz Viktor Arens, *Das Werkmaß in der Baukunst des Mittelalters* (dissertation, Bonn), Würzburg 1938. Arens provides evidence of the appearance of the Roman foot in numerous buildings dating from the Carolingian period up to and into the thirteenth century.
239 Marcus Vitruvius Pollio, *Zehn Bücher über Architektur*, translated and annotated by Curt Fensterbusch, Darmstadt 1987[4], pp. 202-211.
240 Arens 1938, p. 74 ff.
241 W. Schirmer, *architectura* 24, 1994, Figs. 2, 3, and especially 6, p. 188 ff.
242 The dimensions in black were calculated by Professor Marcel Erné of the Mathematical Institute of the Technical University of Hannover.
243 W. Schirmer, 'Castel del Monte', in: *Akademie-Journal* 1/94, p. 18, right column, p. 19, Fig. 5, and p. 20, Fig. 8.
244 Matthias Untermann, *Der Zentralbau im Mittelalter*, Darmstadt 1989.
245 Exceptions are several castles of the Teutonic Order in Prussia (e.g., Rehden, Heilsberg), for which a connection with the architecture of Frederick II cannot be excluded.

246 See H. R. Hahnloser, *Villard d'Honnecourt. Kritische Gesamtausgabe des Bauhüttenbuches* (Ms. fr. 19093 of the Bibliothèque Nationale de Paris), Vienna 1935.

247 Matthias Untermann 1989, especially p. 72 ff.

248 M. S. Bulatov, *Geometric Harmonization in the Architecture of Central Asia in the Ninth and Fifteenth Centuries* (in Russian), 2nd edn., Moscow 1988.

249 Max Koecher, 'Castel del Monte und das Oktogon', in: *Miscellanea Mathematica* 1991, pp. 221–223.

250 See Matthias Untermann 1989, p. 34 ff., and Günther Binding, "Geometricis et aritmeticis', Zur mittelalterlichen Bauvermessung', in: *Jahrbuch der Rheinischen Denkmalpflege* 30–31 (1985), pp. 9–24. See also François Bucher, 'Design in Gothic Architecture. A Preliminary Assessment', in: *Journal of the Society of Architectural Historians* 27 (1968), p. 70.

251 The idea that the complex, geometrically conceived ground plan could have been realized without design drawings and without transfer of the design — or essential parts of it — to the building site is not reconstructible. Even if no traces of such markings can any longer be found at the site, this does not speak against their existence at the start of building. Ropes and posts can be removed without leaving any traces. Such a regular symmetrical design cannot possibly have been erected 'freehand'. If Wulf Schirmer writes, in his report in 'Federico II, immagine e potere', Venice 1994, p. 285 f. (especially p. 286), that it is improbable that the geometric design was marked out prior to the erection of the building due to the great unevenness of the terrain, the question of how such a complex design could have been realized without fixed points remains unanswered. The rocky ground surely constituted a firm foundation, but the unevenness made points of orientation necessary, established using geometric instruments (see note 197). Everything we know about the construction of a foundation prior to the erection of a building during that period (see Robert Mark, ed.: *Vom Fundament zum Dachgewölbe, Großbauten und ihre Konstruktion von der Antike bis zur Renaissance*. Basel/Berlin/Boston 1995, original English edition Cambridge, Mass. 1993, p. 20 ff., especially p. 27) indicates that great importance was attached to the creation of the foundation. It may be presumed that the building was begun with the large main octagon, which is very regular and displays convincing units of the Roman foot measure in its extensions (this volume, p. 170 and passim). W. Schirmer concedes (*architectura* 24, 1994, pp. 185–197) "that the geometric figures proposed by H. Götze could have formed the basis for the elaboration of a design drawing or sketch (!)" (note 15 on p. 193). How should one envisage this, however, especially when one reads in the text on p. 193 "that the marking out of the wall alignments of the basic octagon as well as the doors and the walls dividing the rooms was done step by step as construction progressed." It is hardly conceivable that such a clearly defined design would not have determined the building progress from the start, but would instead have been realized from one construction phase to the next. The walls dividing the rooms are not applicable as an argument against the basic geometric design, which concentrated on the three homeomorphic octagons and is apparent in them. The interior construction is secondary in comparison. W. Schirmer contradicts himself concerning the successive marking out of the wall alignments (p. 193) when he writes on p. 195 "that with regard to wall thicknesses and room measurements, the building was planned and laid out (!) in its finished dimensions...." On p. 200 f. he concludes again "that the ideal form of the tower-fortified building element and the significance [sic!] of the number eight and the octagon formed the starting point for the design certainly may not be doubted." I have taken this view and proven it geometrically here; it is part of the architectural-historical interpretation. However, when he concludes (p. 201) that Castel del Monte "above and beyond the significance of the tower-fortified building element and the number 8... primarily [is] a program that fulfills itself step by step with its construction, [a] program and a compromise at the same time," he leaves everything uncertain. What is this program, other than the geometric design whose existence W. Schirmer only hesitantly accepts? This design defines, and at the same time represents, the 'significance' of the building!

252 This volume, p. 115 ff.

253 This volume, p 119, Fig. 166 c, note 136; Kurt Weitzmann, *Spätantike und frühchristliche Buchmalerei*, Munich 1977, p. 60 f., Fig. 15 on p. 60.

254 Janine Lancha, *Mosaiques Géométriques, Les Ateliers de Vienne-Isère*, Rome 1977, pp. 139 and 149, Figs. 75 and 75 to Inv. No. 164; F. Artaud, *Mosaiques de Lyon et des Départments Méridionaux de la France*, Paris 1818, plate XXIX.

255 See note 249 above.

256 Gunther Wolf, 'Die Wiener Reichskrone', in: *Schriften des Kunsthistorischen Museums Wien* 1 (1995).

257 See note 246 above.

258 Cf. Gülru Necipoğlu: *The Topkapi Scroll — Geometry and Ornament in Islamic Architecture*, Santa Monica 1995 (further cited as GN/TS), p. 68, right column.

259 See Heinz Götze (further abbreviated as HG): 'Castel del Monte, Entwurf und Ausführung', in: *Daidalos* 1996, p. 58.

260 *Veröffentlichung des Institutes für Geschichte der arabisch-islamische Wissenschaften* (edited by Fuat Sezgin), series B, reprints dept. Instrumentengrunde, vol. 3 (to be continued), Frankfurt. a. M 1991.
261 Averages of all measurements, not including the shortened east-west distance due to the portal side.
262 HG 1996, p. 58.
263 For Castel del Monte as a monohedron (see p. 113) the ground plan is repeated on each story (with the exception of constructional variations, e.g., wall thicknesses).
264 Cf. G. Downey, 'Byzantine Architects, Their Training and Methods', in: *Byzantion* 18, 1948, pp. 99–118, especially p. 106ff.
265 Octagons or eight-pointed stars occur in the Topkapi Scroll more than 60 times!
266 GN/TS, p. 92: "... from where it spread to the west between the 12th and 13th centuries, when it experienced its full flowering" (note 7); p. 102, right column.
267 See above GN/TS, p. 79, right column, and p. 103.
268 See GN/TS, p. 140, left column, and note 24: Nasir al-Din al-Tusi, thirteenth century. This corresponds to the statement made by Jean Mignot at the conference of master builders for the cathedral of Milan, 1391/92: "Ars sine scientia nihil est."
269 See Yvonne Dold-Samplonius: 'How al-Kashi Measures the Muqarnas: A second look', in: *Mathematische Probleme im Mittelalter*, Wiesbaden 1996, pp. 57–90.
270 Cf. also GN/TS, p. 44, right column.
271 See GN/TS, pp. 22ff. and 68ff.
272 GN/TS, p. 23.
273 André Paccard, *Le Maroc et l'Artisanat Traditionelle Islamique dans l'Architecture*, 4th edn., Annecy 1983[4]; English translation: 'Traditional Islamic Craft in Moroccan Architecture', 1980
274 GN/TS p. 23, middle and right columns, note 57.
275 GN/TS, pp. 24 and 25.
276 See note 269.
277 Cf. D. Sack, 1996, p. 41ff.
278 See U. Haarmann, *Das Pyramidenbuch des Abu Ga far al-Idrisi*, Beirut 1991, p. 65 of the Arabic text, which H. Halm of Tübingen kindly translated for me: the envoy, whose name is illegible, "came to the pyramids and inspected the inscriptions in various characters, then copied down several lines; he was asked what it concerned, and he said it was Latin and translated it into Arabic."
279 See p. 106ff. of this volume, note 111.
280 Cf. here also GN/TS, p. 101, right column, and note 25. In my opinion, the interactions and connections between Muslim and Gothic architecture in the Middle Ages have not been sufficiently examined.
281 Here we can be fairly sure that the amazingly exact results of measurements were already being achieved with the help of geodetic instruments.
282 See H. M. Schaller, 1993, p. 132.
283 Plotin I, 43.31–33: 'Über das Schöne', translated into German by J. W. von Goethe (*Zahme Xenien III*, in: *Goethes Werke in 10 Bänden*, Artemis edition, vol. 1, Zurich 1961, p. 629).
284 Gunther Wolf: 'Die Wiener Reichskrone', in: *Schriften des Kunsthistorischen Museums Wien*, 1 (1995).
285 *Die Kunstdenkmäler der Stadt Aachen*, Düsseldorf 1981 (reprint), 3 vols. (Die Kunstdenkmäler der Rheinprovinz, vol. 10), vol. 1, Das Münster zu Aachen, p. 41, Fig. 18, p. 76, Fig. 30, plates I and VI. This building as well is based on the eight-pointed star as the design figure. The points of the star lie in the corners of the alignments of the imagined octagon of the outer walls, which are flattened to a decahexagon. The distribution of the ambulatory arches corresponds to those of the mausoleum of Qubbat aṣ-Ṣulaibīya in Sāmarrā (p. 122, this volume, Fig. 171). Following the style of the Pfalzkapelle in Aachen stands the octagonal centrally planned abbey stemming from the Ottonian period which later became the collegiate church of Ottmarsheim, Haut-Rhin. It was founded in 1030 by the Hapsburg Rudolf von Altenburg and consecrated in 1049 by Pope Leo IX, from the family of the counts of Egisheim. Except for the transformation of the outer walls of the Aachen octagon into a decahexagon, the ground plan of Ottmarsheim is completely identical to that of the Pfalzkapelle — including the placement of the arches in the ambulatory (Figs. 272 and 273, this volume).
286 The consecration inscription on this chandelier reads: "Frederick, Catholic emperor of the Roman Empire, pledged — with the bidding to heed the fact that number and form unite harmoniously with the dimensions of this magnificent temple — this crown of lights — octagonal — as a princely gift...." It is striking that the octagonal form is called attention to expressly. It is probable that dimension and number were pointed out in reference to the dedicatory inscription on the octagon itself, which reads: "If the living stones are combined in peace to form a unit, if *number and dimension* correspond in every part, thus will shine the work of the lord who created the hall...." (translation after Jacob Kremer, *Der Aachener Dom, Zeugnis und Zeichen*, Mönchengladbach 1979, pp. 18 and 24). Cf. further Walter Maas, *Der Aachener Dom*, Cologne 1984, pp. 77–79.
287 See H. M. Schaller, 1993, p. 131.
288 Although the comparison of Castel del Monte to the New Temple of Jerusalem appears plausible to me on the basis of the principal measurement of 100 royal ells for a side of the circumscribed square, I cannot agree on all points with the further calculations. I have been able to show that the geometrically deter-

mined ground plan of Castel del Monte is based on the formula $a : 2a : (\sqrt{2}-1)$. The plan thus consists of a clearly definable geometric system, the individual dimensions of which are connected with and all dependent on one another. In other words: when a definite original dimension is given — in this case the length of a side of the circumscribed square — then the other principal dimensions are firmly defined and determined *consequential* dimensions on the basis of the geometric connection; therefore, they cannot be determined arbitrarily. This is also quite evident from Huber's (1996) Figs. 2 and 3 (pp. 124/125): the superimposed grid cannot be described as "shaping the entire ground plan." Incidentally, the length of the side of the initial square, given as 5183 cm, is not really so clear; if one measures from west to east on W. Schirmer's ground plan, the result is 5186 cm, from north to south 5194 cm. On p. 126 f. Huber shows a circumscribed square whose sides are defined by the inner (not visible) sides of the octagon. However, the sides of the octagon that meet the main wall are ca. 15 cm longer than all other sides. Moreover, there is no agreement of the lines of this "inner" square with the line of any actual wall of the building. This does not negate the fundamental correctness of Huber's proposal! It also does not necessarily contradict my suggestion about the existence of the Roman foot (see above, p. 169). Mr. Huber has outlined the connection between the royal ell and other important traditional units of measure, such as the Roman foot (Huber 1993, p. 112, Fig. 46).

289   H. M. Schaller, 1993, p. 138.
290   Cf. Florian Huber, 1996.
291   H. M. Schaller, 1974, p. 109 f.
292   G. Wolf, 'Überlegungen zum Gründungsdatum von Castel del Monte', in: *Kaiser Friedrich II: – Manuskripte und Veröffentlichungen*, pp. 113–116. Heidelberg 1997.

Selected Bibliography
and List of the Literature Cited in Abbreviated Form

Aachen: *Die Kunstdenkmäler der Stadt Aachen*, 3 vols. (Die Kunstdenkmäler der Rheinprovinz, vol. 10), Düsseldorf 1981 (reprint)

Abdul-Hak, M. Selim: 'La reconstitution d'une partie de Qaṣr al-Ḥair al-Ġarbī au Musée de Damas' (in Arabic), in: *Les Annales Archéologiques de Syrie*, 1B, Damascus 1951, pp. 5–57, plates 15–24

Abulafia, David: 'Kantorowicz and Frederick II', in: Davis, H. C. (ed.), *The Journal of the Historical Association*, vol. 62, no. 204, London 1977, pp. 193–210

Abulafia, David: 'Kantorowicz and Frederick II' in: Abulafia, D., *Italy, Sicily and the Mediterranean 1100–1400*, London 1987

Abulafia, David: *Frederick II, A Medieval Emperor*, London 1988

Abulafia, David: *Herrscher zwischen den Kulturen*, Berlin 1991

Ackermann, J. S.: "'Ars sine scientia nihil est.' Gothic Theory of Architecture at the Cathedral of Milan", in: *The Art Bulletin*, vol. XXXI, no. 1, March 1949, pp. 84–111

Adam, Ernst: *Die Baukunst der Stauferzeit in Baden-Württemberg und im Elsaß*, Stuttgart/Aalen 1977

Agnello, Giuseppe: *L'Architettura Sveva in Sicilia*, Collezione Meridionale, Rome 1935 (reprint Syracuse 1986)

Agnello, Guido: 'Il Castello svevo di Prato', in: *Rinasa*, III, 1954, pp. 147–227

Agnello, Guido: 'Problemi ed Aspetti dell'Architettura Sveva', in: *Palladio*, vol. X, 1960, pp. 37–47

Ahrens, Dieter/Rottländer, Rolf C. A. (eds.): *Ordo et Mensura*, I. Interdisziplinärer Kongreß für Historische Metrologie, from September 7 to 10, 1989 in the Städtisches Museum Simeonstift Trier, Sankt Katharinen (unpublished manuscript)

de Angelis d'Ossat, Gioacchino: 'Sugli edifici ottogonali a cupola nell'Antichità e nel Medioevo', in: *Atti del I° Congresso Nazionale di Storia dell'Architettura*, Florence 1938

Arborio Mella, Frederico A.: *Gli Arabi e l'Islam, Storia, Civiltà, Cultura*, Milan 1981

Arens, Fritz Viktor: *Das Werkmaß in der Baukunst des Mittelalters* (Dissertation, Bonn), Würzburg 1938

Artaud, F.: *Mosaiques de Lyon et des Départments Méridionaux de la France*, Paris 1818, plate XXIX

Badawy, Alexander: *Ancient Egyptian Architectural Design*, University of California Publications Near Eastern Studies, vol. 4, Berkeley/Los Angeles 1965

Bandmann, G.: *Mittelalterliche Architektur als Bedeutungsträger*, Berlin 1951

Barry, L.: 'Notes sur les ruines de Yonga', in: *Bulletin Archéologique du Comité des Travaux Historiques et Scientifiques*, Paris 1885, pp. 320–324

Bauer F. L.: 'Sternpolygone und Hyperwürfel' in: *Miscellanea Mathematica*, Berlin/Heidelberg 1991, pp. 7–43

Beijing 1985: *Historic Chinese Architecture*, compiled by the Department of Architecture, Qinghua University, Beijing 1985

Bertaux, Émile: 'Castel del Monte et les architects français de l'empereur Frédéric II' in: *Comptes-rendus des séances de L'Académie des Inscriptions et Belles Lettres*, Paris, August 1897, p. 432 ff.

Bertaux, Émile: *L'Art dans l'Italie méridionale de la fin de l'Empire romain à la conquête de Charles d'Anjou*, 2 vols., Paris 1904

Bertaux, Émile: *L'Art dans L'Italie méridionale*, Aggiornamento dell'Opera di Emile Bertaux sotto la Direzione di Adriano Prandi, 6 vols., vols. 1–3, Rome 1969, vols. 4–6, Rome 1978

Binding, Günther: 'Geometricis et aritmeticis, Zur mittelalterlichen Bauvermessung', in: *Jahrbuch der Rheinischen Denkmalpflege 30–31*, 1985

Bologna Ferdinando: '"Cesaris Imperio Regni Custodia Fio". La Porta di Capua e la "Interpretatio Imperialis"', in: *Nel Segno di Federico II, Atti del IV Convegno Internazionale di Studi della Fondazione Napoli Novantanove*, Naples 1989, pp. 159–189 with figs.

Bottari, Stefano: *Monumenti svevi di Sicilia*, Palermo 1950

Bottari, Stefano: 'Ancora su le Origini dei Castelli Svevi della Sicilia', in: *Atti del Convegno Internazionale di Studi Federiciani in Palermo 1950*, Palermo 1952, pp. 501–505

Boureau, Alain: *Histoires d'un Historien. Kantorowicz*, Paris 1990

Brentjes, Burchard: *Kunst des Islam. Mittelasien*, Leipzig 1979 (2nd edn. 1982)

Brentjes, Burchard: 'Das "Ur-Mandala"(?) von Daschly-3', in: *Iranica Antiqua*, vol. 18, Leiden 1983

Brentjes, Burchard/Bretanizki, L./Weimarn, B.: *Die Kunst Aserbaidschans vom 4. bis zum 8. Jahrhundert*, Leipzig/Weinheim 1988

Brion, Marcel: *Frédéric II de Hohenstaufen*, Paris 1978

Bucher, François: 'Design in Gothic Architecture. A Preliminary Assessment', in: *Journal of the Society of Architectural Historians*, vol. XXVII, March 1968, pp. 49–71

Bulatov, M. S.: *The Mausoleum of the Samanids. Pearl of Central Asian Architecture* (in Russian), Taschkent 1976

Bulatov, M. S.: *Geometric Harmonization in the Architecture of Central Asia in the Ninth and Fifteenth Centuries* (in Russian), Moscow 1978

Burckhardt, Titus: *Die maurische Kultur in Spanien*, Munich 1980

Cadei, Antonio: 'Architettura federiciana. La Questione delle Componenti Islamiche', in: *Nel Segno di Federico II, Atti del IV Convegno Internazionale di Studi della Fondazione Napoli Novantanove*, Naples 1989, pp. 143–158 with figs.

Cantor, Moritz: *Vorlesungen über die Geschichte der Mathematik*, vol. 1, Leipzig 1907 (3rd edn.)

Carandini, A., Ricci, A., de Vos, M.: *Filosofiana, The Villa of Piazza Armerina, Sicily*, Palermo 1982

Carcani, Cajetanus (ed.): *Constitutiones regum regni utriusque Siciliae...*, Naples 1786

Carlier, Patricia: *Qastal, château umayyade?*, Aix-Marseille 1984

Chen, Doron: 'The Design of the Dome of the Rock in Jerusalem', in: *Palestine Exploration Quarterly*, 112, 1980, pp. 41–50

Chierici, Gino: 'Castel del Monte', in: *I Monumenti Italiani, rilievi, raccolti a cura della Reale Accademia d'Italia*, no. 1, Rome 1934

Claussen, Peter Cornelius: 'Die Statue Friedrich II. vom Brückentor in Capua (1234–1239)', in: *Festschrift für Hartmut Biermann*, Weinheim 1990, pp. 19–39

Conway, John H. / Guy, Richard K.: *The Book of Numbers*, New York 1996, p. 172, Figs. 6–8

Cordaro, Michele: 'La porta di Capua', in: *Annuario dell'Istituto di Storia dell'Arte dell'Università di Roma*, Rome 1974–1976

Cormack, Patrick: *Castles of Britain*, New York 1982

Creswell, Keppel Archibald Cameron: *A Short Account of Early Muslim Architecture*, Harmondsworth 1958

Daffa', 'Alī 'Abd Allāh/Stroyls, John J.: *Studies in the Exact Sciences in Medieval Islam*, New York 1984

Daneshvari, Abbas: *Medieval Tomb Towers of Iran*, Lexington 1986

Degeorge, Gérard: *Syrie*, Paris 1983

Deichmann, Friedrich Wilhelm: 'Die Bärtigen mit dem Lorbeerkranz vom Brückentor Friedrich II. zu Capua', in: Deuchler, Florens (ed.): *Von Angesicht zu Angesicht, Michael Stettler zum 70. Geburtstag*, Bern 1983

Diehl, C.: *L'Afrique byzantine I*, Paris 1896

Diehl, Ch.: *Manuel d'art byzantin*, Paris 1925/26

Dodds, Jerrilynn D. (ed): *Al-Andalus, The Art of Islamic Spain*, The Metropolitan Museum of Art, New York 1992, pp. 326–327, no. 92, with two figures

Dold-Samplonius, Yvonne: 'The Volume of Domes in Arabic Mathematics', in: *Vestigia Mathematica. Studies in medieval and early modern mathematics in honour of H. L. L. Busard*, Amsterdam 1993, p. 93 ff.

Dold-Samplonius, Yvonne: 'How al-Kashi Measures the Muqarnas: A second look', in: *Mathematische Probleme im Mittelalter*, Wiesbaden 1996, pp. 57–90

Dölger, Franz Josef: 'Zur Symbolik des altchristlichen Taufhauses. Das Oktogon und die Symbolik der Achtzahl', in: *Antike und Christentum*, Münster/Westf. 1933

Downey, G.: 'Byzantine Architects, Their Training and Methods', in: *Byzantion, Revue Internationale des Études Byzantines*, vol. 18, 1948, pp. 99–118

Ebbinghaus, H.-D. et al.: 'Zahlen', in: *Grundwissen Mathematik I*, Berlin/Heidelberg/New York 1988

El-Said, Issem/Parman, Ayşe: *Geometric Concepts in Islamic Art*, London 1976

Endres, Franz Carl/Schimmel, Annemarie: *Das Mysterium der Zahl*. Diederichs Gelbe Reihe (2nd edn.), Cologne 1985

Erdmann, Kurt and Hanns: *Das anatolische Karavansaray des 13. Jahrhunderts*, part 3: 'Die Ornamente', Berlin 1976

Fibonacci, Leonardo: *Le livre des nombres carrés*, translated and introduced by Paul ver Eecke, Bruges/Paris 1952

Finley, M. L./Smith, Denis Mack/Duggan, Christopher: *A History of Sicily*, London 1986

Fischer, Klaus/Jansen, Michael/Pieper, Jan: *Architektur des indischen Subkontinents*, Darmstadt 1987

Flüeler, Nikolaus (ed.): *Das große Burgenbuch der Schweiz*, Munich 1977

Franciosi, Filippo: 'L'Irrazionalità nella Matematica Greca Arcaica', in: *Bollettino dell'Istituto di Filologia Greca dell'Università di Padova*, Suppl. 2, Rome 1977

Franco, Saro/Cultraro, Massimo: *Guida attraverso i saloni del Castello Normanno di Adranò*, Adranò (no date)

Frank, Josef: 'Die Verwendung des Astrolabs nach al-Chwaritzmi', in: *Abhandlungen zur Geschichte der Naturwissenschaften und der Medizin*, no. III, Erlangen 1922

Frankl, Paul: 'The Secret of the Medieval Masons', with Panofsky, Erwin: 'An Explanation of Stornaloco's Formula', in: *The Art Bulletin*, vol. XXVII, no. 1, March 1945, pp. 46-64

Friedman, David: *Florentine New Towns. Urban Design in the late Middle Ages*, Cambridge, Mass./London 1988

Friedrich II.: *Über die Kunst, mit Vögeln zu jagen*, in cooperation with Dagmar Odenthal, translated and edited by Carl Arnold Willemsen, 2 vols., Frankfurt a.M. 1964; annotation volume by Carl Arnold Willemsen, Frankfurt a.M. 1970

Gabrieli, Francesco: 'Friedrich II. und die Kultur des Islam', in: *Stupor Mundi. Zur Geschichte Friedrichs II. von Hohenstaufen*, edited by Gunther Wolf, Darmstadt 1982 (2nd edn.)

Gabrieli, Francesco: 'Federico II e il Mondo dell'Islam', in: *Nel Segno di Federico II, Atti del IV Convegno Internazionale di Studi della Fondazione Napoli Novantanove*, Naples 1989, pp. 129-139

Gandz, Salomon: 'Die Harpedonapten oder Seilspanner und Seilknüpfer', in: *Quellen und Studien zur Geschichte der Mathematik, Astronomie und Physik*, sect. B: Studien vol. 1, Berlin 1931

Gericke, Helmuth: *Mathematik in Antike und Orient*, Berlin/Heidelberg/New York/Tokyo 1984

Gernet, Jacques: *Le Monde Chinois*, Paris 1972 (German edn., *Die chinesische Welt*, Frankfurt a.M. 1979)

Giuffrè, Maria: *Castelli e luoghi forti di Sicilia (XII–XVII secolo)*, Palermo 1980

Götze, Heinz: 'Die Baugeometrie von Castel del Monte', in: *Sitzungsbericht der Heidelberger Akademie der Wissenschaft*, Heidelberg 1991

Götze, Heinz: 'Castel del Monte, Entwurf und Ausführung', in: *Daidalos*, March 1996, pp. 53-59

Gombrich, Ernst H.: *Il senso dell'ordine*, Turin 1984

Grabar, Oleg: *The Formation of the Islamic Art*, New Haven/London 1987

Gruyon, Etienne / Stanley, H. Eugene (eds.): *Fractal Forms*, Amsterdam 1991

Guarini, Guarino: *Architettura civile del padre di Guarino Guarini ... opera postuma ...*, Turin 1737

Gurrieri, Francesco (ed.): *Il Castello dell'Imperatore a Prato*, Prato 1975

Haarmann, U.: *Das Pyramidenbuch des Abu Ga far al-Idrisi*, Beirut 1991

Hahn, Hanno: *Die frühe Kirchenbaukunst der Zisterzienser*, Berlin 1957

Hahn, Hanno/Renger-Patzsch, Albert: *Hohenstaufenburgen in Süditalien*, Ingelheim am Rhein 1961

Hahnloser, H.R.: *Villard d'Honnecourt: Kritische Gesamtausgabe des Bauhüttenbuches*, Ms. fr. 19093 of the National Library of Paris, Vienna 1935

Hamilton, R.W.: *Khirbat al-Mafjar*, Oxford 1959

Hammer-Purgstall, Josef: *Codices Arabicos Pericos Túrcicos*, Vienna 1812

Hampe, K.: *Deutsche Kaisergeschichte in der Zeit der Salier und Staufer*, Heidelberg 1949 (10th edn.)

Haseloff, A.: *Die Bauten der Hohenstaufen in Unteritalien*, text vol. I, Leipzig 1920

Haskins, C.H.: *Studies in the History of Mediaeval Science*, Harvard 1927

Hecht, Konrad: 'Maß und Zahl in der gotischen Baukunst', in: *Abhandlungen in der Braunschweigischen Wissenschaftlichen Gesellschaft*, Hildesheim/New York 1979

Heinisch, Klaus J. (ed.): *Kaiser Friedrich II. in Briefen und Berichten seiner Zeit*, Darmstadt 1978 (6th edn.)

Hell, G. / Otto, J.: 'Photogrammetrische Arbeiten an Castel del Monte', in: *Allgemeine Vermessungsnachrichten* 4/92, pp. 169–174

Hinz, Hermann: *Motte und Donjon. Zur Frühgeschichte der mittelalterlichen Adelsburg (Zeitschrift für Archäologie des Mittelalters*, Suppl. I), Cologne/Bonn 1981

Hoag, John D.: *Islam*, Stuttgart 1986

Horst, Eberhard: *Friedrich der Staufer*, Düsseldorf 1975

Hotz, Walter: *Pfalzen und Burgen der Stauferzeit. Geschichte und Gestalt*, Darmstadt 1981

Huber, Florian: 'Angewandte Metrologie in Geschichte und Gegenwart', Trierer Museumsseminare 1993, p. 112, Fig. 46

Huber, Florian: 'Das Castel del Monte Kaiser Friedrichs II. in Apulien als Paradigma für exakte mittelalterliche Bauvermessung', in: *Genauigkeit und Präzision in der Geschichte der Wissenschaften und des Alltags*, PTB Texts, vol. 4, Braunschweig 1996, p. 121 ff.

Huillard-Bréholles, Jean-Louis-Alphonse: *Historia Diplomatica Friderici Secundi sive Constitutiones...*, vols. 1–5, Paris 1852–1861

Jones, Owen: *The Grammar of Ornament*, folio edition, London 1856

Junecke, Hans: 'Die Meßfigur', in: *Jahrbuch des Deutschen Archäologischen Instituts*, Archäologischer Anzeiger, no. 4, 1970

Junecke, Hans: *Die wohlbemessene Ordnung*, Berlin 1982

Kantorowicz, Ernst H.: *Kaiser Friedrich der Zweite*, 2 vols., Berlin 1927; Supplementary volume *Quellennachweise und Exkurse*, 1931 (reprint 1963)

Kaschnitz-Weinberg, Guido: *Mitteilungen des Deutschen Archäologischen Instituts*, Roman section, vol. 60/61, 1953/54, pp. 1-21 and vol. 62, 1955, pp. 1-52

Khan, Alund Nabi: *Multan – History and Architecture*, Institute of Islamic History, Culture and Civilization, Islamic University, Islamabad 1983

King, David A.: 'Astronomie im mittelalterlichen Jemen', in: *Jemen*, edited by Werner Daum, Innsbruck/Frankfurt a.M. 1987

Klein, U. / Zick, W.: 'Castel del Monte – der geodätische Beitrag zur ersten präzisen Bauaufnahme', in: *Allgemeine Vermessungsnachrichten* 4/92, pp. 163–169

Koecher, M.: 'Castel del Monte und das Oktogon', in: *Miscellanea Mathematica*, Berlin/Heidelberg 1991, pp. 221–233

Koepf, H.: *Die gotischen Planrisse der Wiener Sammlungen*, Vienna/Cologne/Graz 1969

Körte, Werner: 'Das Kastell Kaiser Friedrich II. in Lucera', in: *25 Jahre Kaiser-Wilhelm-Gesellschaft zur Förderung der Wissenschaften*, vol. III: *Die Geisteswissenschaften*, Berlin 1937

Kottmann, Albrecht: *Maßverhältnisse in Zisterzienserbauten*, Munich 1968

Kottmann, Albrecht: *Fünftausend Jahre Messen und Bauen*, Stuttgart 1981

Kramers, J.H./Wiet, G. (eds.): *Configuration de la terre*, Beirut/Paris 1964

Krautheimer, Richard: 'Introduction to an Iconography of Medieval Architecture', in: *Journal of the Warburg and Courtauld Institutes*, vol. V, 1942

Kremer, Jacob: *Der Aachener Dom, Zeugnis und Zeichen*, Mönchengladbach 1979

Krönig, Wolfgang: 'Prefazione' zu G. Agnello: *L'Architettura Sveva in Sicilia*, Rome 1935 (reprint Syracuse 1986)

Krönig, Wolfgang: 'Staufische Baukunst in Unteritalien', in: *Beiträge zur Kunst des Mittelalters*, contributions to the First German Congress of Art Historians at Schloß Brühl 1948, Berlin 1950

Krönig, Wolfgang, in: *L'Art dans l'Italie méridionale*, Aggiornamento dell'Opera di Émile Bertaux sotto la direzione di Adriano Prandi, vol. 5, Rome 1978

Krönig, Wolfgang: 'Castel del Monte. Der Bau Friedrichs II.' in: *Intellectual Life at the Court of Frederick II Hohenstaufen. Proceedings of the Symposium*, Center for Advanced Study in the Visual Arts, Washington D.C. 1991

Kubach, Hans Erich: 'Romanik', in: *Weltgeschichte der Architektur*, Stuttgart 1986

Kubler, George/Soria, Martin Sebastian: *Art and Architecture in Spain and Portugal and their American Dominions 1500–1800*, Harmondsworth 1959

Kühnel, Ernst: 'Islamische Einflüsse in der romanischen Kunst', in: *Sitzungs-Berichte der Kunstgeschichtlichen Gesellschaft Berlin*, Nov. 11, 1927

Kühnel, Ernst: 'Das Rautenmotiv an romanischen Fassaden in Italien', in: Rohde, G., et al. (eds.): *Edwin Redslob zum 70. Geburtstag*, Berlin 1955

Lancha, Janine: *Mosaiques Géométriques, Les Ateliers de Vienne (Isère)*, Rome 1977, pp. 139 and 149, Figs. 75 and 76 to Inv. No. 164

Lee, A.J.: 'Islamic Star Patterns', in: *Muqarnas. An Annual on Islamic Art and Architecture*, vol. 4, Leiden 1987, pp. 182–197

Lehmann-Brockhaus, Otto: *Abruzzen und Molise: Kunst und Geschichte*, Munich 1983

Leistikow, D.: 'Burgen und Schlösser der Capitanata im 13. Jahrhundert. Ein Überblick', in: *Bonner Jahrbücher*, CLXXI, 1971, pp. 416–441

Leistikow, Dankwart: 'Zum Mandat Kaiser Friedrichs II. von 1240 für Castel del Monte', in: *Deutsches Archiv für Erforschung des Mittelalters*, 50, 1994

Lewis, B.: *The World of Islam*, London 1976

Lézine, A.: *Le Ribat de Sousse. Suivi de Notes sur le Ribat de Monastir*, Tunis 1956

List, K.: 'Die Wasserburg Lahr', in: *Burgen und Schlösser, Zeitschrift für Burgenkunde*, 1970

López de Arenas, Diego: *La Carpinteria de lo Blanco*, Lectura dibujada del primer manuscrito de Diego López de Arenas par Enrique Nuere, Ministero de Cultura, Madrid 1985

Lord Robert, Howard: *The Origins of the War of 1870*, Cambridge 1924

Lowry, Glenn D.: 'Humayun's Tomb: Form, Function and Meaning in Early Mughal Architecture', in: *Muqarnas. An Annual on Islamic Art and Architecture*, vol. 4, Leiden 1987, pp. 133–148

Maas, Walter: *Der Aachener Dom*, Cologne 1984, pp. 77–79

Marconi, Pirro: 'Note sull'Ariete del Museo di Palermo', in: *Bollettino d'Arte del Ministerio dell'E.N.*, September 1930, pp. 132–143

Mark, Robert (ed.): *Architectural Technology up to the Scientific Revolution. The Art and Structure of Large-scale Buildings*, Cambridge, Mass. 1993. German translation: *Vom Fundament zum Deckengewölbe, Großbauten und ihre Konstruktion von der Antike bis zur Renaissance*, Basel/Berlin/Boston 1995

Meckseper, Cord: 'Castel del Monte. Seine Voraussetzungen in der nordwesteuropäischen Baukunst', in: *Zeitschrift für Kunstgeschichte*, vol. XXXIII, 1970, pp. 211–231

Meckseper, Cord: 'Über die Fünfeckkonstruktion bei Villard de Honnecourt und im späten Mittelalter', in: *Architectura. Zeitschrift für Geschichte der Baukunst*, vol. 13 I, Munich 1983

Merra, E.: *Castel del Monte*, Molfetta 1964 (3rd edn.)

Mertens, Dieter: 'Castellum oder Ribāṭ? Das Küstenfort in Selinunt', in: *Mitteilungen des Deutschen Archäologischen Instituts Istanbul*, vol. 39, 1989, pp. 391–398

Meyer, Heinz/Suntrup, Rudolf: *Lexikon der mittelalterlichen Zahlenbedeutungen*, Munich 1987, columns 565–580. Detailed bibliography in column 580

Michell, George: *Architecture of the Islamic World*, London 1978

Mola, Riccardo: *Restauri in Puglia (1971–1983) II*, Fasano (Brindisi) 1983

Molajoli, B.: *Castel del Monte*, Naples 1958

Montesinos-Amilibia, J.M.: *Classical Tesselations and Three-Manifolds*, Berlin/Heidelberg 1987

Moscati, Sabatino: *Die Karthager*, Stuttgart/Zurich 1984

Müller, Edith: *Gruppentheoretische und strukturanalytische Untersuchungen der maurischen Ornamente aus der Alhambra in Granada* (Dissertation) Zurich 1944

Muratori: *Ryccardi de Sancto Germano Notarii Cronica*, Muratori, Rer. Ital. Script. Nova Ser. vol. VII, Paris II (A cura di C. A. Garufi), 1938

Murray, Peter: *Die Architektur der Renaissance in Italien*, Stuttgart 1980

Musca, Giosuè: *Castel del Monte, Il Reale e l'Immaginario*, Bari 1981

Nasr, Seyyed Hossein: *An Introduction to Islamic Cosmological Doctrines*, Cambridge/Mass. 1964

Necipoğlu, Gülru: *The Topkapi Scroll – Geometry and Ornament in Islamic Architecture*, Santa Monica 1995

Northedge, Alastair/Faulkner, Robin: 'Survey Season at Sāmarrā 1986', in: *Iraq*, 49/1987, pp. 143–173

d'Onofrio, M.: 'La Torre Cilindrica di Caserta Vecchia', in: *Napoli Nobilissima*, vol. VII, no. 1, 1969

Paccard, André: *Le Maroc et l'artisanat traditionnel islamique dans l'architecture*, 2 vols. (4th edn.), Annecy 1983

Paolini, P.: 'Nuovi aspetti sul castel Maniace di Siracusa', in: *Atti del III congresso di architettura fortificata (1981)*, Rome 1985, pp. 215–222

Papadopoulo, Alexandre: *Islamische Kunst*, Freiburg/Basel/Vienna 1977

Papadopoulo, Alexandre: *Islam and Muslim Art*, London 1980

Passanti, Mario: *Nel Mondo Magico di Guarino Guarini*, Turin 1963

Pedretti, Carlo: *Leonardo da Vinci, Architekt*, Stuttgart/Zurich 1981

Petronotis, Argyres: *Bauritzlinien und andere Aufschnürungen am Unterbau griechischer Bauwerke in der Archaik und Klassik. Eine Studie zur Baukunst und -technik der Hellenen* (published by the author), Munich 1968

Pevsner, Nikolaus/Honour, Hugh/Fleming, John: *Lexikon der Weltarchitektur*, Munich 1987 (2nd edn.)

Pugačenkova, Galina Anatol'evna: *Puti razvitija architektury Južnogo Turkmenistana pory rabovladenija i feodalizma [The Paths of Development of the Architecture of Southern Turkmenistan During Slavery and Feudalism]*, Moscow 1958

Pugačenkova, Galina Anatol'evna: *Chefs-d'oeuvre d'architecture de L'Asie centrale XIV–XV siècle*, Paris 1981

Propyläen-Kunstgeschichte: *Propyläen-Kunstgeschichte*, Berlin, vol. 4 (1973), vol. 6 (1972), vol. 8 (1970), vol. 9 (1970)

de Renzi, Salvatore: *Storia documentata della scuola medica di Salerno*, Milan 1967 (new edition)

Repertorio: *Architettura sveva nell'Italia meridionale, Repertorio dei Castelli Federiciani*, Palazzo Pretorio, Prato, March–September 1975 (A cura di Arnaldo Bruschi e Gaetano Miarelli Mariani), Florence 1975

Revue Archéologique: *Revue Archéologique*, third series, vol. XII, 1888, pp. 14–23

Romanini, Angiola Maria (ed.): *Federico II e l'arte del Duecento Italiano, Atti della III settimana di studi di storia dell'arte medievale dell'Università di Roma*, 2 vols., Lecce 1980

Roshdi, Rashed: 'Fibonacci et les Mathématiques Arabes', in: *Actes du Colloque sur Fédéric II. et les Savoirs*, Palermo 1990

Roshdi, Rashed: 'Fibonacci et les Mathématiques Arabes', in: *Micrologus II, Le scienze alla corte di Federico II*, pp. 145–160. Paris 1994

Rosiny, Tonny: *Pakistan*, Cologne 1983

Rottländer, Rolf C. A.: *Antike Längenmaße. Untersuchungen über ihre Zusammenhänge*, Braunschweig/Wiesbaden 1979

Rottländer, Rolf C. A./Heinz, Werner/Neumaier, Wilfried: 'Untersuchungen am Turm der Winde in Athen', in: *Jahreshefte des Österreichischen Archäologischen Instituts*, vol. LIX, Vienna 1989, pp. 55–92

Sack, Dorothée / Schirmer, Wulf: 'Castel del Monte', in: *Koldewey Gesellschaft. Bericht über die 37. Tagung für Ausgrabungswissenschaft und Bauforschung, May 27–31, 1992 in Duderstadt*, pp. 84–91

Sack, Dorothée: 'Castel del Monte e l'Oriente', in: *Federico II, immagine e potere*. Exhibition catalogue, Bari 1995, pp. 295–303

Sack, Dorothée: 'Einige Bemerkungen zu den Beziehungen zwischen Castel del Monte und dem Orient', in: *Kunst im Reich Kaiser Friedrichs II. von Hohenstaufen*, Munich/Berlin 1996, p. 41 ff.

Saponaro, Giorgio (ed.): *Castel del Monte*, Bari 1981

Saumagne, M. C.: 'Antiquités de la région de Sfax', in: *Bulletin Archéologique du Comité des Travaux Historiques et Scientifiques*, vol. 34/35, Paris 1938, pp. 750–763

Schaller, Hans Martin: 'Kaiser Friedrich II., Verwandler der Welt', in: *Persönlichkeit und Geschichte*, vol. 34, Göttingen 1971 (2nd edn.)

Schaller, Hans Martin: 'Die Kaiseridee Friedrich II.', in: *Probleme um Friedrich II., Vorträge und Forschungen*, edited by Konstanzer Arbeitskreis für mittelalterliche Geschichte, vol. XVI, Sigmaringen 1974

Schaller, Hans Martin: 'Die Staufer in Apulien', in: *Staufisches Apulien. Schriften zur staufischen Geschichte und Kunst*, vol. 13, Göppingen 1993

Schaller, Hans Martin: 'Die Frömmigkeit Friedrichs II', in: *Schriften zur staufischen Geschichte und Kunst*, vol. 15, Göppingen 1996, pp. 128–151

Schaller, Hans Martin: 'Die Herrschaftszeichen Kaiser Friedrichs II', in: *Schriften zur staufischen Geschichte und Kunst*, vol. 16, Göppingen 1997, pp. 58–105

Schirmer, Wulf: 'Castel del Monte', in: *Akademie-Journal* 1/94, p. 19, Fig. 5, and p. 20, Fig. 8

Schirmer, Wulf with G. Hell, U. Hess, D. Sack and W. Zick: 'Castel del Monte. Neue Forschungen zur Architektur Friedrichs II', in: *architectura* 1/2, 1994, pp. 185–224

Schirmer, Wulf: 'Castel del Monte: Osservazioni sull'edificio', in: *Federico II, immagine e potere*. Exhibition catalogue, Bari 1995, pp. 285–294

Schirmer, Wulf / Sack, Dorothée: 'Castel del Monte', in: *Kunst im Reich Kaiser Friedrichs II. von Hohenstaufen*, Munich/Berlin 1996, pp. 35–44

Schmidt, Anna Maria: 'La Fortezza di Mazzallaccar' in: *Bollettino d'Arte*, series V, vol. 57, 1972, pp. 90–93

Schneider, Gerd: *Geometrische Bauornamente der Seldschuken in Kleinasien*, Wiesbaden 1980

Schweitzer, Bernhard: *J. G. Herders 'Plastik' und die Entstehung der neueren Kunstwissenschaft*, Leipzig 1948

Seckel, Dietrich: 'Taigenkyû, das Heiligtum des Yuiitsu-Shintô. Eine Studie zur Symbolik und Geschichte der japanischen Architektur', in: *Monumenta Nipponica*, 6, Tokyo 1943, pp. 52–85

Sezgin, Fuat: *Geschichte des Arabischen Schrifttums*, vol. V, Leiden 1974

Sezgin, Fuat (ed.): *Veröffentlichung des Institutes für Geschichte der arabisch-islamische Wissenschaften*, series B, Frankfurt a.M. 1991

Shearer, Creswell: *The Renaissance of Architecture in Southern Italy. A Study of Frederic II of Hohenstaufen and the Capua Triumphator Archway and Towers*, Cambridge 1935

Shelby, Lon R.: 'The Geometrical Knowledge of Medieval Master Masons', in: *Speculum, Journal of Medieval Studies*, vol. XLVII, Cambridge/Mass. 1972

Shelby, Lon R.: *Gothic Design Techniques. The Fifteenth Century Design Booklet of Mathes Roriczer and Hanns Schmuttermayer*, London/Amsterdam 1977

von Simson, Otto: *Die gotische Kathedrale*, Darmstadt 1982 (4th edn.)

Soergel, Gerda: *Untersuchungen über den theoretischen Architekturentwurf von 1450-1550 in Italien*, Munich 1958

de Solla Price, Derek: 'Geometrical and Scientific Talismans and Symbolisms', in: *Science since Babylon*, New Haven/London 1976

de Solla Price, Derek J./Nobel, John V.: 'The Water Clock in the Tower of the Winds', in: *American Journal of Archaeology*, 1968

Speiser, Andreas: *Die mathematische Denkweise (Wissenschaft und Kultur)*, Basel 1945 (2nd edn.)

Staats, Reinhart: *Theologie der Reichskrone: Ottonische 'Renovatio imperii' im Spiegel einer Insignie*, Monographien zur Geschichte des Mittelalters, vol. 13, Stuttgart 1976

di Stefano, Guido: *Monumenti della Sicilia Normanna*, a cura di W. Krönig, Palermo 1979 (2nd edn.)

Stern, Henri: 'Notes sur l'architecture des châteaux omeyyades', in: *Ars Islamica* (The Department of Fine Arts), vols. XI–XII, University of Michigan Press, Ann Arbor 1946

Sthamer, Eduard: *Dokumente zur Geschichte der Kastellbauten Kaiser Friedrichs II. und Karls I. von Anjou*, parts I and II, Leipzig 1912, 1926

Stierlin, Henri: *Architektur des Islam vom Atlantik zum Ganges*, Zurich/Freiburg 1979

Strika, Vincenzo: 'The turbah of Zumurrud Khatun in Baghdad', in: *Annali dell'Istituto Orientale di Napoli*, vol. XXXVIII (N. S. XXVIII), 1978, pp. 283–296

Stupor Mundi: *Stupor Mundi, Zur Geschichte Friedrichs II. von Hohenstaufen*, edited by Gunther Wolf, Darmstadt 1982 (2nd edn)

Stürner, Wolfgang: *Friedrich II, Teil I: Die Königsherrschaft in Sizilien und Deutschland 1194–1220*, Darmstadt 1992

Suter, Heinrich: 'Beiträge zur Geschichte der Mathematik bei den Griechen und Arabern', in: *Abhandlungen zur Geschichte der Naturwissenschaften und der Medizin*, edited by Oskar Schulz, no. IV, Erlangen 1922

Suter, Heinrich: 'Beiträge zu den Beziehungen Kaiser Friedrichs II. zu den zeitgenössischen Gelehrten des Ostens und Westens, insbesondere zu dem arabischen Enzyklopädisten Kamāl ad-Dīn ibn Yūnis', in: *Abhandlungen zur Geschichte der Naturwissenschaften und der Medizin*, no. IV, Erlangen 1922

Swinburne, E.: *Viaggio nelle due Sicilie negli anni 1777 e 1778*

Talbi, Mohamed: *L'Emirat aghlabide*, Paris 1966

Taploo, Rita: 'Octagon in Islamic Tombs — a Structured Exigency or a Metaphysical Symbolism' in: *Islamic Culture*, II/2, 1977, pp. 141–149

Testi Christiani, M. L.: *Nicola Pisano architetto e scultore*, Pisa 1987

Thiersch, August: *Die Proportionen in der Architektur (Handbuch der Architektur, IV/1)*, Leipzig 1904

Tits, Jacques: 'Symmetrie' in: *Miscellanea Mathematica*, Berlin/Heidelberg 1991, pp. 293–304

Torraca, Francesco: *Studi su la lirica italiana del duecento*, Bologna 1902

Torraca, Francesco: 'Le origini. L'età sveva', in: *Storia della Università di Napoli*, edited by Gennaro Maria Monti, Naples 1924

'Toscana und Apulien, Beiträge zum Problemkreis der Herkunft des Nicola Pisano', in: *Zeitschrift für Kunstgeschichte*, edited by Wolfgang Krönig, vol. 16, 1953, pp. 78–82, 101–144, 222–233

Trier 1990: *Karolingische Handschriften, Beda-Handschriften und Frühdrucke*, exhibition catalogue, edited by the Kulturstiftung der Länder and the Stadtbibliothek Trier, 1990

Underwood, Paul A.: 'The Fountain of Life in Manuscripts of the Gospels', in: *Dumbarton Oaks Papers*, no. 5, Cambridge/Mass. 1950

Untermann, Mathias: *Der Zentralbau im Mittelalter*, Darmstadt 1989

Valensise, Marina: 'Storici e Storia. Ernst Kantorowicz', in: *Rivista Storica Italiana*, 1989, pp. 195–221

Violett-le-Duc, E.: *Entretiens sur l'Architecture*, Paris 1863

de Vita, Raffaele (ed.): *Castelli, torri ed opere fortificate di Puglia*, Bari 1982 (2nd edn.)

Vitruvius, Marcus Pollio: *Zehn Bücher über Architektur*, translated and annotated by Curt Fensterbusch (2nd edn.), Darmstadt 1987

Vogel, Kurt: 'Fibonacci, Leonardo, or Leonardo of Pisa', in: *Dictionary of Scientific Biography*, edited by Charles Coulston Gillispie, vol. IV, New York 1971, pp. 604–613

van der Waerden, B. L.: *Geometry and Algebra in Ancient Civilisation*, Berlin/Heidelberg 1983

van der Waerden, B. L.: *A History of Algebra*, Berlin/Heidelberg 1985

Wagner-Rieger, Renate: *Die italienische Baukunst zu Beginn der Gotik*, part II: Süd- und Mittelitalien, Graz/Cologne 1957 (Publikationen des Österreichischen Kulturinstitutes in Rom, Abteilung für Historische Studien, sect. I, vol. 2, parts 1 and 2)

Weitzmann, Kurt: *Spätantike und frühchristliche Buchmalerei*, Munich 1977

Weyl, Hermann: 'Symmetrie', in: *Wissenschaft und Kultur*, vol. II, Boston/Basel/Stuttgart 1981 (2nd edn.)

Wiedemann, Eilhard: 'Miszellen. Fragen aus dem Gebiet der Naturwissenschaften, gestellt von Friedrich II., dem Hohenstaufer', in: *Archiv für Kulturgeschichte*, edited by W. Goetz/G. Steinhausen, vol. XI, no. 4, Leipzig/Berlin 1914, pp. 483–485

Willemsen, Carl Arnold: *Kaiser Friedrichs II. Triumphtor zu Capua. Ein Denkmal staufischer Kunst in Süditalien*, Wiesbaden 1953

Willemsen, Carl Arnold: 'Die Bauten der Hohenstaufer in Süditalien. Neue Grabungs- und Forschungsergebnisse', in: *Arbeitsgemeinschaft für Forschung des Landes Nordrhein-Westfalen*, no. 149, issue 25, Cologne/Opladen 1968

Willemsen, Carl Arnold: *Apulien, Land der Normannen, Land der Staufen*, Cologne 1968 (3rd edn.)

Willemsen, Carl Arnold: *Apulien, Kathedralen und Kastelle. Ein Kunstführer durch das normannisch-staufische Apulien*, Cologne 1973 (2nd edn.)

Willemsen, Carl Arnold: 'Die Bauten der Staufer', in: *Die Zeit der Staufer*, vol. 3, exhibition catalogue, Stuttgart 1977

Willemsen, Carl Arnold: *Castel del Monte*, Insel-Bücherei no. 619, Frankfurt a.M. 1982 (2nd edn.)

Willemsen, Carl Arnold: *Castel del Monte*, Bari 1984

Wilsdorf, Christian: 'Le Burgstall de Guebwiller, Les châteaux octogonaux d'Alsace et les constructions de l'empereur Frédéric II', in: *Annuaire de la Société d'Histoire des régions de Thann-Guebwiller*, 1970–1972

Woepcke, M. F.: 'Recherches sur l'histoire des sciences mathémathiques chez les Orientaux d'après des traités inédits Arabes et Persans', in: *Journal Asiatique*, series 5, vol. V, Paris 1855

Wolf, Gunther G.: 'Kaiser Friedrich II. und das Recht', in: *Zeitschrift der Savigny Stiftung für Rechtsgeschichte*, RA 102, 1985

Wolf, Gunther G.: 'Die Wiener Reichskrone', in: *Schriften des Kunsthistorischen Museums*, vol. 1, Vienna 1995

Wolf, Gunther G.: *Satura Mediaevalis*, vol. III: Stauferzeit, Heidelberg 1996

Wolf, Karl Lothar/Wolff, Robert: *Symmetrie*, Munich/Cologne 1956

Wolf, Reinhart: *Castillos, Burgen in Spanien*, Munich 1983

Wollin, Nils G.: *Desprez en Italie*, Malmö 1935

Yakov, Malkiel: 'Storiografia e Miti Culturali: Federico II e Ernst Kantorowicz', in: *Nel Segno di Federico II, Atti del IV Convegno Internazionale di Studi della Fondazione Napoli Novantanove*, Naples 1989, pp. 43–63

Zhuoyun, Yu (ed.): *Palaces of the Forbidden City*, Hong Kong 1984

Sources of Illustrations

*Groups of illustrations that stem from one source are listed first, in alphabetical order.
These are followed by individual sources, listed according to the number of the figure in the book.*

Agnello, Giuseppe: *L'Archittectura Sveva in Sicilia*, Rome 1935: 23 (p. 39, Fig. 11), 25 (p. 47, Fig. 16), 34 (p. 75, Fig. 43), 35 (p. 64, Fig. 32), 47 (p. 172, Fig. 114), 48 (p. 422, Fig. 272), 113 (p. 358, Fig. 233), 114 (p. 354, Fig. 230), 117 (p. 375, Fig. 259).

Alinari, Rome: 7 (no. 19536), 8 (no. 19524), 11 (no. 19543).

Papadopoulo, Alexandre: *Islamische Kunst*, Freiburg/Basel/Vienna, 1977: 178 (Fig. 1127), 181 (Fig. 1127), 187 (Fig. 1004), 201 (Fig. 1090).

Propyläen Kunstgeschichte, Berlin 1967–1973: 17 (vol. 1, 1967, p. 292, Fig. 61), 18 (vol. 1, 1967, p. 293, Fig. 62), 46 (vol. 1, 1967, p. 256, Fig. 43 F), 58 (vol. 4, 1973, p. 155, Fig. 13), 61 (vol. 4, 1973, p. 156, Fig. 5), 64 (vol. 4, 1973, p. 167, Fig. 8), 66 (vol. 4, 1973, p. 170), 67 (vol. 4, 1973, p. 225, Fig. 49), 95 (vol. 4, 1973, plate XI), 124 (vol. 8, 1970, p. 370, Fig. 32), 165 (vol. 5, 1969, p. 216, Fig. 40).

Renger-Patzsch, Albert, from: Hahn, Hanno and Renger-Patzsch, Albert: *Hohenstaufenburgen in Süditalien*, Ingelheim am Rhein 1961: 5 (plate 95), 14 (p. 20, Fig. 5), 15 (p. 20, Fig. 4), 16 (p. 21, Fig. 9), 19 (plate 3), 52 (plate 67), 80 (p. 37, Fig. 28), 96 (plate 73).

Stern, Henri: 'Notes sur l'architecture des châteaux omeyyades', in: *Ars Islamica* (The Department of Fine Arts), vols. XI–XII, University of Michigan Press, Ann Arbor 1946: 62 (Fig. 10), 63 (Fig. 5).

Stierlin, Henri: *Architektur des Islam vom Atlantik zum Ganges*, Zurich/Freiburg 1979: 57 (Fig. p. 51), 59 (p. 52, Fig. 17), 60 (Fig. p. 55), 65 (Fig. p. 54), 76 (Fig. p. 162), 77 (Fig. p. 221), 223 (Fig. p. 127).

Willemsen, Carl A.: *Das Falkenbuch Kaiser Friedrichs II., De arte venandi cum avibus*, zwölf Faksimile-Drucke aus dem Codex Ms. Palatinus Latinus 1071 der Biblioteca Apostolica Vaticana, Graz 1973: 13 (p. 29 oben/fol. IV).

— : 'Die Bauten Kaiser Friedrichs II. in Süditalien', in: *Die Zeit der Staufer*, vol. III, Stuttgart 1977: 83, 85 (p. 157, Figs. 14, 15), 93 (p. 158, Fig. 17), 99 (plate xxxvi, Fig. 44), 102 (p. 154, Fig. 11).

— : *Kaiser Friedrichs II. Triumphtor zu Capua*, Wiesbaden 1953: 94 (Fig. 105), 97 (Fig. 3). For Figs. 83, 85, 97, and 99 original prints or films were kindly provided.

Wollin, Nils G.: *Desprez en Italie*, Malmö 1935: 75 (p. 205, Fig. 45), 84 (p. 202, Fig. 38), 86 (p. 203, Fig. 39), 89 (p. 204, Fig. 42), 112 (p. 238, Fig. 113).

Illustrations taken from pictures by Franz Schlechter, Heidelberg (1991): Frontispiece, Figs. 22, 24, 26, 27, 32, 36–45, 134–140, 142–149, 156, 157, 225, 248, 249.

Illustrations taken from my own pictures or sketches: p. 6/7, p. 8, p. 17, Figs. 12, 20, 21, 29, 30, 33, 49–51, 53, 71–74, 78, 79, 81, 82, 88, 90, 91, 98, 100, 101, 103–105, 107, 109, 115, 116, 118, 125, 128–131, 133, 141, 150, 151, 153, 155, 158, 161, 162, 166f, 167, 168, 174, 176, 177, 185, 188, 192, 202, 203, 205, 207, 213, 215–217, 220, 226, 228–235, 244–247, 250, 251, 252, 255–260, 262, 263, 265, 266, 268, 270, 273, 276.

Individual sources:

1: Deutsches Archäologisches Institut, Römische Abteilung (Inst. Neg. no. 541 105).
2: Foto Dehnert, Göppingen.
3: Biblioteca Apostolica Vaticana, Vatican City.
4: Claussen, Peter Cornelius: 'Die Statue Friedrichs II. vom Brückentor in Capua (1234–1239)', in: *Festschrift für Hartmut Biermann*, Weinheim 1990 (p. 300, Fig. 18).
6: Schmidt, Wieland, *Die Manessische Handschrift*, Berlin 1977 (p. 7, plate I).
9, 10: Kunsthistorisches Institut der Universität Wien (Si 6/2, Si 2/1).
28: Mirabelli, Ubaldo: *Nella luce di Palermo*, Palermo 1982 (Fig. p. 25).
31: Soprintendenza alle Antichità, Palermo, Gabinetto Fotografico (Lastro 16267, NT 8365).
54, 56: Lézine, Alexandre: *Le Ribat de Sousse. Suivi de Notes sur le Ribat de Monastir*, Tunis 1956 (Fig. 14, Fig. XXXIIIb).

55: Editions Kahia, Tunis.
68: Schmidt, Anna Maria: 'La Fortezza di Mazzallaccar', in: *Bollettino d'Arte*, series V, vol. 57, 1972 (p. 91, Fig. 1).
69: Mertens, Dieter: 'Castellum oder Ribat? Das Küstenfort in Selinunt', in: *Mitteilungen des Deutschen Archäologischen Instituts Istanbul*, vol. 39, 1989 (p. 394, Fig. 2).
70: Keystone Press AG, Zurich.
87: Deutsches Archäologisches Institut, Cairo branch, Rainer Stadelmann, Nairi Hampikian.
92: Mola, Riccardo: *Restauri in Puglia (1971–1983) II*, Fasano (Brindisi) 1983
106, 108: Haseloff, Arthur: *Die Bauten der Hohenstaufen in Unteritalien*, text volume, Leipzig 1920 (p. 360, Fig. 72; p. 358, Fig. 71, 5).
110, 111: Lehmann-Brockhaus, Otto: *Abruzzen und Molise*, Munich 1983 (p. 331, Fig. 59; Fig. 209).
119–121: Reinhard Spehr, Dresden.
122: List, K.: 'Die Wasserburg Lahr', in: *Burgen und Schlösser*, Zeitschrift für Burgenkunde, 1970 (p. 44, Fig. 3a).
123: Meyer, Werner and Widmer, Eduard: *Das große Burgenbuch der Schweiz*, Zurich 1981[4] (Fig. p. 139).
126, 127: Chierici, Gino: 'Castel del Monte', in: *I Monumenti Italiani, rilievi, raccolti a cura della Reale Accademia d'Italia*, no. 1, Rome 1934 (plates I and II).
132: Soc. Trimboli, Pescara.
152: Gruyon, Etienne and Stanley, H. Eugene (eds.): *Fractal Forms*, Amsterdam 1991.
154: Fede, Burgos (no. 10[F]-3).
159: Bauer, F. L.: 'Sternpolygone und Hyperwürfel', in: *Miscellanea Mathematica*, Berlin/Heidelberg 1991 (Fig. p. 12).
160, 209: Hahnloser, Hans R.: *Das Bauhüttenbuch des Villard de Honnecourt*, Vienna 1935 (Figs. 98/99 and plate 41).
163: Rheinisches Bildarchiv, Cologne, RBA 215016.

164, 200: Deichmann, Friedrich Wilhelm: *Frühchristliche Kirchen in Rom*, Basel 1948 (p. 40, plan 8b; p. 25 ff., plan 4).
166a: Lancha, Janine: *Mosaiques Géométriques, Les Ateliers de Vienne (Isère)*, Rome 1977, Figs. 75 and 76.
166b, 218: Carandini, A., Ricci, A., de Vos, M.: *Filosofiana, The Villa of Piazza Armerina, Sicily*, Palermo 1982, (plate VI, Fig. 22; p. 246, Fig. 146 and front cover of volume of plates).
166c: Austrian National Library, Vienna, photo archives.
166d: Lewis, B.: *The World of Islam*, London 1976 (p. 244, Fig. 17).
166e: *Collection of Fine Arts in China*, vol. 1, 'Palace Architecture' (in Chinese), plate 23, Beijing 1987.
169: Brentjes, Burchard: *Kunst des Islam. Mittelasien*, Leipzig 1979, 1982[2] (Fig. p. 24).
170: Alastair Northedge, London 1987.
171, 206: Creswell, K. A. C.: *A Short Account of Early Muslim Architecture*, Harmondsworth 1958, Fig. 59, Fig. 2.
172, 173: Redrawn from Bulatov, M. S.: *The Mausoleum of the Samanids. Pearl of Central Asian Architecture* (in Russian), Taschkent 1976 (p. 339, Fig. 61/199a; p. 185, Fig. 115).
175, 189: Fischer, Klaus/Jansen, Michael/Pieper, Jan: *Architektur des indischen Subkontinents*, Darmstadt 1987 (p. 197, Fig. 209; p. 226, Fig. 252).
179, 182: Witte, P.: Deutsches Archäologisches Institut, Abteilung Madrid (no. R 130-83-3; R 130-76-5).
180: Balestrini, Bruno, in: Hoag, John D.: *Islam*, Stuttgart 1986 (p. 46, Fig. 74).
183: Kubach, Hans Erich: 'Romanik', in: *Weltgeschichte der Architektur*, Stuttgart 1986 (p. 152, Fig. 243).
184: Pugačenkova, G. A.: *Die Wege der architektonischen Ent-*

wicklung Südturkmenistans ... (in Russian), Moscow 1958 (Fig. p. 363).
186: Grécy, Jules: *Die Alhambra zu Granada*, Worms 1990 (p. 38, plate 46).
190: Lee, A. J.: 'Islamic Star Patterns', in: *Muqarnas. An Annual on Islamic Art and Architecture*, vol. 4, Leiden 1987, p. 182ff.
191: Paccard, André: *Le Maroc et l'artisanat traditionnel islamique dans l'architecture*, Annecy 1983[4], vol. I (Fig. p. 210).
193: Murray, Peter: *Die Architektur der Renaissance in Italien*, Stuttgart 1980 (p. 87, Fig. 66).
194: Soergel, Gerda: *Untersuchungen über den theoretischen Architekturentwurf von 1450–1550 in Italien*, Munich 1958 (p. 127, Fig. 23).
195: Passanti, Mario: *Nel mondo magico di Guarino Guarini*, Turin 1963 (p. 133, Fig. 11).
196: Bibliotheca Hertziana, Rome.
197, 198, 199: Kraus, Theodor: 'Das Römische Weltreich', in: *Propyläen Kunstgeschichte*, vol. 2, Berlin 1967 (p. 195, Fig. 35; p. 180, Fig. 19; p. 194, Fig. 34).
204: Werner Neumeister, Munich.
208: Michell, George: *Architecture of the Islamic World*, London 1978 (p. 114, Fig. 7).
210: Germanisches Nationalmuseum, Nuremberg.
211, 212: Shelby, Lon R.: *Gothic Design Techniques. The Fifteenth Century Design Booklet of Mathes Roriczer and Hanns Schmuttermayer*, London/Amsterdam 1977, pp. 85, 86.
214, 221: Bibliothèque Nationale, Paris.
219: Rafael Pérez Gómez, Granada.
222, 261, 264, 267: Necipoğlu, Gülru: *The Topkapi Scroll – Geometry and Ornament in Islamic Architecture*, Santa Monica 1995 (p. 149, Fig. 113b; Fig. 104; Fig. 97; p. 25, Fig. 50).
224: Redrawn from Bulatov, M. S.: *Geometric Harmonization in the*

Sources of Illustrations 233

Architecture of Central Asia in the Ninth to Fifteenth Centuries (in Russian), Moscow 1978, p. 177, Fig. 86.

227: Cantor, Moritz: *Vorlesungen über die Geschichte der Mathematik*, Leipzig 1907³, vol. 1 (p. 106, Fig. 9).

236–243: Marcel Erné, Hannover.

253, 254: Susanne Krömker, Interdisciplinary Center for Scientific Calculation at the University of Heidelberg (Director: Prof. Dr. W. Jäger).

269: Dold-Samplonius, Yvonne: 'How al-Kashi Measures the Muqarnas: A second look', in: *Mathematische Probleme im Mittelalter*, Wiesbaden 1996, p. 71, Fig. 12.

271: Kroll, Gerhard: *Auf den Spuren Jesu*, Stuttgart 1983, p. 54, Fig. 30.

272: Haupt, A., in: Faymonville, Karl: *Das Münster zu Aachen*, Düsseldorf 1916 (Fig. p. 76).

274, 275: Ann Münchow, Aachen.

Index of Names and Places

The index contains common place names and building locations.

Aachen   24, 27
—, Pfalzkapelle   130, 206, *Figs. 272, 274, 275*
'Abd al-Malik, caliph   137
Abū Ya'qūb Yūsuf II, caliph   118, *Fig. 166*
Abū Muslim   122
'Abd ar-Rahmān I, emir of Córdoba   52
Aḏerbaiğān, mausoleum of Şaiḫ Dursun   *Fig. 224*
Aditya   123
Adranò, donjon   30, 62, 79, 80, *Figs. 81, 112*
al-Ağdābīya (Libya), monument to al-Qa'im bi-amr Allāh Abū l-Qāsim   125
Agra (India), Tağ Maḥall   131, *Fig. 189*
Alberti, Leon Battista   143
Aleppo (Syria), Madrasa al-Firdaus   59, *Fig. 76*
Altamura, cathedral   29
Altenburg   87, 88
Ambrose, St.   117
Anagni   104, *Fig. 152*
Ancona   21
Andria   98, 207
'Anğar (Syria), Omayyad city   52, *Fig. 57*
Anicia Juliana, Byzantine princess   *Fig. 166c*
Anjou, Charles of (Angevin)   61, 68
L'Aquila   48
Aquinus, Thomas   26
Archimedes   142
Aristotle   116, *Fig. 161*
Athens, Tower of the Winds   117, *Fig. 162*
Atsiz, Muḥammad   *Fig. 173*

Augusta, citadel   43, 44, 47, 56, 75, *Fig. 47*
Augustine   143

Baġdād (Iraq)   52, 121, *Fig. 169*
Bajae, temple of Diana   135, *Fig. 198*
Barbaro, Daniele   113
Bari   30, 48, 51, 74
Beijing (China), Temple of Heaven   117, 121
—, imperial palace   121, *Fig. 166e*
Bejaïa (Algeria)   26
Belisarius   22
di Benevento, Roffredo   26
Bethlehem   206, 207, 208
—, Church of the Nativity   206, *Fig. 271*
Bodiam Castle   88
Braunschweig   88
Buḫārā (Uzbekistan)   122
—, mausoleum of the Samanid Ismā'īl   122, *Fig. 172*
Burgos (Spain), Las Huelgas cloister   107, *Fig. 154*
Bursa (Turkey), 'Green Tomb'   123
al-Būzağānī, Abū l-Wafā'   153, 154, *Fig. 221*
Byzantium   26, 87, 135

Cairo   63, 64
—, al-Ḥākim mosque   63, *Fig. 87*
Cambridgeshire, Ely cathedral   130
Capua, bridge citadel   19, 37, 67–72, 74, 90, 92, 112, 158, 203, *Figs. 93, 94, 96–98, 155*
Caserta   98
di Caserta, Contessa Violante   74
Caserta Vecchia, residential tower   72–74, 112, *Figs. 100, 101, 155*
Castel Maniace (see Syracuse)

Castel Ursino (see Catania)
Catania, Castel Ursino   44–62, 64, 71, 91, 110, 154, *Figs. 48–52, 73, 155*
Châlons-sur-Marne, Nôtre-Dame   106
Chambord, castle   88, *Fig. 124*
Charlemagne, Holy Roman emperor, king of the Franks   24, 206
Chartres, cathedral   109
Chinard, Philippe   106
Constance of Aragon   21, 27
Córdoba (Spain)   22, 52
—, Omayyad mosque   38, 52, 125–128, 134, 135, 152, 158, *Figs. 178, 179, 181, 182*
da Cortona, Domenico   *Fig. 124*
Cremona, Gerard of   143
Cyrrhestes, Andronicus   117, *Fig. 162*

Damascus (Syria)   52, 71
Dante Alighieri   140
Datong (China), Shanhua temple   121
Delhi (India), mausoleum of Humāyūn   123, 131, 133, 163, 187, *Figs. 187, 188, 192, 255*
—, mausoleum of 'Īsā Ḫān Niyāzī   123
Desprez, Louis Jean   59, 61–64, 80, *Figs. 75, 84, 86, 89, 112*
Dietrich von Meißen, margrave   83, 84, 87, 88
Diocletian   48, 135
Diophantes   153
Dioscurides   186, *Fig. 166c*
Dürer, Albrecht   113

Eberbach, abbey church   169
Edessa (Mesopotamia)   87

Eger (Cheb) 87
Egisheim, donjon 82, 83
Ehrenfels, Christian von 15
Eleusis, telesterion 43, *Fig. 46*
Enna (Castro Giovanni), Torre di Federico 37, 80–83, *Figs. 113–117, 155*
Erfurt 87
Euclid 113, 153, 191, 195

Fano, basilica 170
al-Fārābī Abū Naṣr 153
Fibonacci, Leonardo 13, 26, 58, 147
Florence, San Giovanni *Fig. 165*
Foggia 32, 60, 61
—, palace 208, *Fig. 21*
Francesca, Piero della 113
Frederick I, Barbarossa, German emperor 21, 24, 27, 206
Frederick II, German emperor 12, 19–34, 48, 51, 56, 60, 61, 66, 68, 69, 71, 74, 76, 83, 87, 88, 89, 91, 92, 98, 100, 107, 112, 114, 138, 155, 156, 158, 159, 184, 189, 202, 203, 206–208, *Figs. 1–5, 13, 96*
Frederick III, Count of Leiningen 83

Ġiyāt ad-Dīn al-Kāšī 153
Ġiyāt ad-Dīn Tuġluq 123
Giza 202
Goethe, Johann Wolfgang von 205
Granada (Spain), Alhambra 14, 129, 147, 148, 150, 151, 152, 154, 158, 204, *Figs. 186, 219, 220*
Gregory IX, Pope 24
Guarini, Guarino 113, 134, *Figs. 195, 196*
Gubbio 159
Guebwiller, Burgstall 82
Guiscard, Robert 24

Halle 87
Ḫān Ġahān Tīlanġānī, mausoleum of (India) 123
Heinrich von Morungen 87
Henry VI, German emperor 21, *Fig. 6*
Henry VII, son of Frederick II 208
Herder, Johann Gottfried 111
Hippocrates 23
Hirbat al-Minya (Syria) *Fig. 61*
Honnecourt, Villard de 60, 113f., 140, 141, 143, 145, 161, 184, 189, 203, *Figs. 160, 209*

Honnecourt, Villard de ('Master 2') 60, 141
Hugo, abbot of Murbach 83
Hungary, Andreas of 68
al-Ḫwārizmī, Muḥammad 153

Ibn Ḥauqal 22, 50, 53, 57
Ibn Idrīs 53
Ibn Sīnā, Abū ʿAli al-Husain ibn ʿAbdallāh (Avicenna) 153
Iesi 21
Innocent II, Pope 21
Isabella of Brienne 98, 207
Isabella of England 98, 207
d'Isernia, Andrea 26
Iṣfahān (Iran), Šāh mosque *Fig. 223*
Isidorus of Miletus 113
Istanbul (Turkey) 195
—, Šaihzade mosque 59, *Fig. 77*

Jerusalem 207
—, Dome of the Rock (Qubbat aṣ-Saḫra) 137, 138, 158, 206, *Figs. 204–206*
—, al-Aqṣā mosque 137

al-Kashi 200
Kyoto (Japan), Koryu-ji temple 120, *Fig. 168*

Lagopesole, citadel 74, *Figs. 99, 155*
Lahr, citadel 88, *Fig. 122*
Lamṭa (Tunisia), ribāṭ 51
Lauterberg, abbey 87
Leipzig 88
da Lentini, Riccardo 33
Leonardo da Vinci 113, 116, 133, 135, *Fig. 193*
Lodi 33
Lucera, citadel 35, 58, 60–67, 79, 89, 91, 208, *Figs. 74, 78–80, 83–86, 155*

Maghreb 196, 199
Marburg, church of St. Elizabeth 140
Marches, the 207
Marw (Turkmenistan) 122
—, mausoleum of Sulṭān Sanğar 122, *Fig. 173*
Mazzallaccar, Arab fortress 53, 57, *Fig. 68*
Mazara 50

Mecca (Saudi Arabia) 52
—, Kaʿba 137
Melfi 24
—, citadel 29, *Fig. 15*
Messina 113, 134
Metz 87
Mignot, Jean 140, 143
Mīhne (Turkmenistan), mausoleum of Abū Saʿid 128, *Fig. 184*
Moncada, Conte Giovanni Tomaso 80
Monreale, cathedral 23, *Fig. 12*
Montealegre (Spain), castle 88
Montefusculo, R. de 159
Monte Sant' Angelo, church of Santa Maria Maggiore 64, *Fig. 92*
Montreux, Château Chillon 57, *Fig. 70*
Morges, fortress 88, *Fig. 123*
Morimond, abbey 87
Morocco 132, 200
al-Mšattā (Jordan), Omayyad palace *Fig. 65*
Muḥammad, prophet 52, 122, 137
Muʿāwiya I, caliph of Damascus 52
Multān (Pakistan), mausoleum of ʿAli Akbar 123, *Figs. 176, 177*
—, mausoleum of Rukn-i-ʿĀlam 123, 124, *Fig. 174*
al-Munastīr (Tunisia), ribāṭ 51, 52, *Fig. 56*

Naples 26, 67
Nara (Japan), Horyu-ji temple 129, *Fig. 167*
Naumburg, cathedral 140
Nero 135
Neu-Leiningen, fortress 83

Oria, citadel 29, 51, *Fig. 14*
Oschatz 83, 84
Osterlant 83–88, *Figs. 119–121*
Ottmarsheim, collegiate church *Fig. 273*
Otto IV, emperor 87, 88

Pacioli, Luca 113
Palermo 21–23, 34, 49, 63, 208
—, Capella Palatina 22
—, cathedral 21, 28, *Fig. 5*
—, La Martorana 22, *Fig. 11*
—, San Cataldo 22, *Figs. 8–10*
—, San Giovanni degli Eremiti 22, *Fig. 7*

Pappos 195
Parler, Heinrich 143
Pasargadae (Iran), Achaemenian building 29, *Fig. 17*
Paternò, donjon 30, 62, *Fig. 82*
Persepolis (Iran), Achaemenian building 29, 30, *Fig. 18*
Piazza Armerina, villa *Fig. 166b*
Pisa 26, 63, 64
—, cathedral 64, *Figs. 90, 129*
—, San Nicola 64
—, San Paolo da Ripa d'Arno 64
Pisa, Bonanus of 25
Pisa, Leonardo of (see Fibonacci, Leonardo)
Pisano, Nicola 112
Plato 59, 114, 116, 205, *Figs. 160, 161*
Plotin 205
Polycleitus 114
Prato, citadel ('Castello dell'Imperatore') 47, 74–76, *Figs. 102–104*
Ptolemy 23, 153
Pythagoras 58–60, 155, 160, *Fig. 71*

al-Qā'im bi-amr Allāh Abū l-Qāsim 125
Qairawān (Tunisia) 50
Qal'at Sim'ān (Syria), St. Symeon Stylites 137, 139
Qaṣr al-Ḥair al-Ġarbī (Syria) 71, *Figs. 66, 95*
Qaṣr al-Ḥair aš-Šārqī (Syria) *Fig. 64*
Qaṣr al-Ḥarāna (Jordan) 52, 57, *Figs. 59, 60*
Qaṣr aṭ-Ṭūba (Jordan) *Fig. 63*
al-Qaṣtal (Jordan) *Fig. 62*
Quedlinburg 88

Ravenna, San Vitale 130, 135, 139, 158, 163, *Fig. 202*
Reims, St. Remi 106
Riegl, Alois 111
Rocca di Calascio, citadel 48, 79, 80, *Figs. 110, 111*
Roger I, Norman prince 24, 50

Roger II, Norman king 21, 50, 53
Rome, Domus Aurea 135, *Fig. 201*
—, Lateran *Fig. 164*
—, Pantheon 135, *Fig. 197*
—, Santa Constanza 135, *Fig. 200*
Roriczer, Mathes 114, 141–143, 160, *Figs. 211, 212*

Salerno 23, 26
Salimbene 19
Samarqand (Uzbekistan) 153
—, mausoleum of Šādī Mulk Aqā 128, *Fig. 185*
Sāmarrā (Iraq), mausoleum Qubbat aṣ-Ṣulaibīya 122, 135, 195, *Figs. 170, 171*
Sambuca 53
de Sanctis, Francesco 26
da San Germano, Riccardo 67, 92
San Severino, Riccardo 74
Santa Maria del Monte 159
Sasāram (India), mausoleum of Šīr Šāh Sūr 123
Schmuttermayer, Hanns 114, 141, 143, *Fig. 210*
Scotus, Michael 27
Selinunt, fortress 56, *Fig. 69*
Serlio, Sebastiano 133, 135, *Fig. 194*
Shotoku, Japanese prince 120, 121
Sinan 59
Siponto 59, 64
—, Santa Maria 31, 59, 64, *Figs. 75, 88, 89*
Solari, Tommaso 19, 69, *Fig. 4*
Split (Dalmatia), Diocletian's palace 135, *Fig. 199*
Steinsberg, fortress 82, *Fig. 118*
Sūsa (Tunisia) 50
—, ribāṭ 51, 52, 56, *Figs. 54, 55*
Syracuse 51, 56
—, Castel Maniace 30, 33–44, 47–49, 56–58, 61, 64, 71, 82, 90, 91, 96, 98, 112, 156, *Figs. 22–45, 72, 155, 225*

Tamerlane 131, 196
Termoli, citadel 62, 77–80, *Figs. 106–109, 155*
Toledo (Spain) 23
—, Bāb Mardūm mosque 128
Torres del Rio (Spain), Santo Sepolcro 128, *Fig. 183*
Trani, Barisanus of 25
Trani, harbor fortress 29, *Figs. 16, 19*
Troia, cathedral 31, 64, 156, *Figs. 91, 130*
Tunis (Tunisia) 50, 52
Turin, San Lorenzo 134, 135, *Figs. 195, 196*

Uč (Pakistan), tomb of Bībī Ġāwandī 124, *Fig. 175*
Uluǧ Beg 153
Usais (Syria), palace of caliph al-Walīd I 52, *Fig. 58*

Valenzano, Ognissanti di Cuti 30, *Fig. 20*
Vanvitelli, Luigi 98
Venice 26
Vienna 186
Vienne-Isère 186, *Fig. 166a*
da Vinci (see Leonardo)
Vitruvius Pollio 59, 170

al-Walid I, caliph 52, *Fig. 58*
Walkenried, cloister 87
William II, Norman king, king of Sicily 23, 25
Wölfflin, Heinrich 111
Würzburg 87

Yingxian (China), Jingtu temple 121

Zaragoza (Spain), La Aljafería 52, 126, *Figs. 67, 180*